Probabilistic Logic Networks

A Comprehensive Framework for Uncertain Inference

Probabilistic Logic Networks

A Comprehensive Framework for Uncertain Inference

Ben Goertzel
Matthew Iklé
Izabela Freire Goertzel
Ari Heljakka

Authors
Ben Goertzel
Novamente LLC
1405 Bernerd Place
Rockville, MD 20851
ben@goertzel.org

Matthew Iklé
Department of Chemistry, Computer Science, and Mathematics
Adams State College
Alamosa, CO 81102
moikle@adams.edu

Izabela Freire Goertzel
Novamente LLC
1405 Bernerd Place
Rockville, MD 20851
izabela@goertzel.org

Ari Heljakka
Novamente LLC
1405 Bernerd Place
Rockville, MD 20851
heljakka@iki.fi

ISBN: 978-1-4419-4578-5 e-ISBN: 978-0-387-76872-4
DOI: 10.1007/978-0-387-76872-4

Printed on acid-free paper

springer.com

Dedicated to the memory of Jeff Pressing

Physicist, psychologist, musician, composer, athlete, polyglot and so much more --

Jeff was one of the most brilliant, fascinating, multidimensional and fully alive human beings any of us will ever know. He was killed in his sleep by a sudden, freak meningitis infection in 2002, while still young and in perfect health, and while in the early stages of co-developing the approach to probabilistic reasoning described in this book.

Jeff saw nearly none of the words of this book and perhaps 25% of the equations. We considered including him as a posthumous coauthor, but decided against this because many of the approaches and ideas we introduced after his death are somewhat radical and we can't be sure he would have approved them. Instead we included him as co-author on the two chapters to whose material he directly contributed. But nonetheless, there are many ways in which the overall PLN theory presented here – with its combination of innovation, formality and practicality -- embodies Jeff's "spirit" as an intellect and as a human being. Jeff, we miss you in so many ways!

Contents

Chapter 1: Introduction

Abstract In this chapter we provide an overview of probabilistic logic networks (PLN), including our motivations for developing PLN and the guiding principles underlying PLN. We discuss foundational choices we made, introduce PLN knowledge representation, and briefly introduce inference rules and truth-values. We also place PLN in context with other approaches to uncertain inference.

1.1 Motivations

This book presents Probabilistic Logic Networks (PLN), a systematic and pragmatic framework for computationally carrying out uncertain reasoning – reasoning about uncertain data, and/or reasoning involving uncertain conclusions. We begin with a few comments about why we believe this is such an interesting and important domain of investigation.

First of all, we hold to a philosophical perspective in which "reasoning" – properly understood – plays a central role in cognitive activity. We realize that other perspectives exist; in particular, logical reasoning is sometimes construed as a special kind of cognition that humans carry out only occasionally, as a deviation from their usual (intuitive, emotional, pragmatic, sensorimotor, etc.) modes of thought. However, we consider this alternative view to be valid only according to a very limited definition of "logic." Construed properly, we suggest, logical reasoning may be understood as the basic framework underlying all forms of cognition, including those conventionally thought of as illogical and irrational. The key to this kind of extended understanding of logic, we argue, is the formulation of an appropriately general theory of uncertain reasoning – where what is meant by the latter phrase is: reasoning based on uncertain knowledge, and/or reasoning leading to uncertain conclusions (whether from certain or uncertain knowledge). Moving from certain to uncertain reasoning opens up a Pandora's box of possibilities, including the ability to encompass within logic things such as induction, abduction, analogy and speculation, and reasoning about time and causality.

While not necessarily pertinent to the technical details of PLN, it is perhaps worth noting that the authors' main focus in exploring uncertain inference has been its pertinence to our broader work on artificial general intelligence (AGI). As elaborated in (Goertzel and Pennachin 2007; Goertzel and Wang 2007; Wang et al 2008), AGI refers to the construction of intelligent systems that can carry out a variety of complex goals in complex environments, based on a rich contextual understanding of themselves, their tasks and their environments. AGI was the original motivating goal of theAI research field, but at the moment it is one among multiple streams of AI research, living alongside other subfields focused on more narrow and specialized problem-solving. One viable approach to achieving powerful AGI, we believe, is to create integrative software systems with uncertain inference at their core. Specifically, PLN has been developed within the context of a larger artificial intelligence project, the Novamente Cognition Engine or NCE (Goertzel 2006), which seeks to achieve general forms of cognition by integrating PLN with several other processes. Recently, the NCE has spawned an open-source sister project called OpenCog, as well (Hart and Goertzel 2008). In the final two chapters we will briefly discuss the implementation of PLN within the NCE, and give a few relevant details of the NCE architecture. However, the vast majority of the discussion of PLN here is independent of the utilization of PLN as a component of the NCE. PLN stands as a conceptual and mathematical construct in its own right, with potential usefulness in a wide variety of AI and AGI applications.

We also feel that the mathematical and conceptual aspects of PLN have the potential to be useful outside the AI context, both as purely mathematical content and as guidance for understanding the nature of probabilistic inference in humans and other natural intelligences. These aspects are not emphasized here but we may address them more thoroughly in future works.

Of course, there is nothing new about the idea that uncertain inference is broadly important and relevant to AI and other domains. Over the past few decades a number of lines of research have been pursued, aimed at formalizing uncertain inference in a manner capable of application across the broad scope of varieties of cognition. PLN incorporates ideas derived from many of these other lines of inquiry, including standard ones like Bayesian probability theory (Jaynes, 2003), fuzzy logic (Zadeh 1989), and less standard ones like the theory of imprecise probabilities (Walley 1991), term logic (Sommers and Englebretsen 2000), Pei Wang's Non-Axiomatic Reasoning System (NARS) (Wang 1996), and algorithmic information theory (Chaitin 1987). For various reasons, which will come out as the book proceeds, we have found each of these prior attempts (and other ones, from which we have not seen fit to appropriate ideas, some of which we will mention below) unsatisfactory as a holistic approach to uncertain inference or as a guide to the creation of an uncertain inference component for use in integrative AGI systems.

Among the general high-level requirements underlying the development of PLN have been the following:

- To enable uncertainty-savvy versions of all known varieties of logical reasoning; including, for instance, higher-order reasoning involving quantifiers, higher-order functions, and so forth.
- To reduce to crisp "theorem prover" style behavior in the limiting case where uncertainty tends to zero.
- To encompass inductive and abductive as well as deductive reasoning.

- To agree with probability theory in those reasoning cases where probability theory, in its current state of development, provides solutions within reasonable calculational effort based on assumptions that are plausible in the context of real-world embodied software systems.
- To gracefully incorporate heuristics not explicitly based on probability theory, in cases where probability theory, at its current state of development, does not provide adequate pragmatic solutions.
- To provide "scalable" reasoning, in the sense of being able to carry out inferences involving at least billions of premises. Of course, when the number of premises is fewer, more intensive and accurate reasoning may be carried out.
- To easily accept input from, and send input to, natural language processing software systems.

The practical application of PLN is still at an early stage. Based on our evidence so far, however, we have found PLN to fulfill the above requirements adequately well, and our intuition remains that it will be found to do so in general. We stress, however, that PLN is an evolving framework, consisting of a conceptual core fleshed out by a heterogeneous combination of components. As PLN applications continue to be developed, it seems likely that various PLN components will be further refined and perhaps some of them replaced entirely. We have found the current component parts of PLN acceptable for our work so far, but we have also frequently been aware of more sophisticated alternative approaches to various subproblems (some drawn from the literature, and some our own inventions), and have avoided pursuing many of such due to a desire for initial simplicity.

The overall structure of PLN theory is relatively simple, and may be described as follows. First, PLN involves some important choices regarding knowledge representation, which lead to specific "schematic forms" for logical inference rules. The knowledge representation may be thought of as a definition of a set of "logical term types" and "logical relationship types," leading to a novel way of graphically modeling bodies of knowledge. It is this graphical interpretation of PLN knowledge representation that led to the "network" part of the name "Probabilistic Logic Networks." It is worth noting that the networks used to recognize knowledge in PLN are weighted directed hypergraphs (Bollobas 1998) much more general than, for example, the binary directed acyclic graphs used in Bayesian network theory (Pearl 1988).

Next, PLN involves specific mathematical formulas for calculating the probability value of the conclusion of an inference rule based on the probability values of the premises plus (in some cases) appropriate background assumptions. It also involves a particular approach to estimating the confidence values with which these probability values are held (weight of evidence, or second-order uncertainty). Finally, the implementation of PLN in software requires important choices regarding the structural representation of inference rules, and also regarding "inference control" – the strategies required to decide what inferences to do in what order, in each particular practical situation.

1.1.1 Why Probability Theory?

In the next few sections of this Introduction we review the conceptual foundations of PLN in a little more detail, beginning with the question: Why choose probability theory as a foundation for the "uncertain" part of uncertain inference?

We note that while probability theory is the foundation of PLN, not all aspects of PLN are based strictly on probability theory. The mathematics of probability theory (and its interconnection with other aspects of mathematics) has not yet been developed to the point where it is feasible to use explicitly probabilistic methods to handle every aspect of the uncertain inference process. Some researchers have reacted to this situation by disregarding probability theory altogether and introducing different conceptual foundations for uncertain inference, such as Dempster-Shafer theory (Dempster 1968; Shafer 1976), Pei Wang's Non-Axiomatic Reasoning System (Wang 1996), possibility theory (Zadeh 1978) and fuzzy set theory (Zadeh 1965). Others have reacted by working within a rigidly probabilistic framework, but limiting the scope of their work fairly severely based on the limitations of the available probabilistic mathematics, avoiding venturing into the more ambiguous domain of creating heuristics oriented toward making probabilistic inference more scalable and pragmatically applicable (this, for instance, is how we would characterize the mainstream work in probabilistic logic as summarized in Hailperin 1996; more comments on this below). Finally, a third reaction – and the one PLN embodies – is to create reasoning systems based on a probabilistic foundation and then layer non-probabilistic ideas on top of this foundation when this is the most convenient way to arrive at useful practical results.

Our faith in probability theory as the ultimately "right" way to quantify uncertainty lies not only in the demonstrated practical applications of probability theory to date, but also in Cox's (1961) axiomatic development of probability theory and ensuing refinements (Hardy 2002), and associated mathematical arguments due to de Finetti (1937) and others. These theorists have shown that if one makes some very basic assumptions about the nature of uncertainty quantification, the rules of elementary probability theory emerge as if by magic. In this section we briefly review these ideas, as they form a key part of the conceptual foundation of the PLN framework.

Cox's original demonstration involved describing a number of properties that should commonsensically hold for any quantification of the "plausibility" of a proposition, and then showing that these properties imply that plausibility must be a scaled version of conventional probability. The properties he specified are, in particular[1],

[1] The following list of properties is paraphrased from the Wikipedia entry for "Cox's Theorem."

1. The plausibility of a proposition determines the plausibility of the proposition's negation; either decreases as the other increases. Because "a double negative is an affirmative," this becomes a functional equation

$$f(f(x)) = x$$

saying that the function f that maps the probability of a proposition to the probability of the proposition's negation is an involution; i.e., it is its own inverse.

2. The plausibility of the conjunction [A & B] of two propositions A, B, depends only on the plausibility of B and that of A ***given*** that B is true. (From this Cox eventually infers that multiplication of probabilities is associative, and then that it may as well be ordinary multiplication of real numbers.) Because of the associative nature of the "and" operation in propositional logic, this becomes a functional equation saying that the function g such that

$$P(A \text{ and } B) = g(P(A), P(B|A))$$

is an associative binary operation. All strictly increasing associative binary operations on the real numbers are isomorphic to multiplication of numbers in the interval [0, 1]. This function therefore may be taken to be multiplication.

3. Suppose [A & B] is equivalent to [C & D]. If we acquire new information A and then acquire further new information B, and update all probabilities each time, the updated probabilities will be the same as if we had first acquired new information C and then acquired further new information D. In view of the fact that multiplication of probabilities can be taken to be ordinary multiplication of real numbers, this becomes a functional equation

$$yf\left(\frac{f(z)}{y}\right) = zf\left(\frac{f(y)}{z}\right)$$

where f is as above.

Cox's theorem states, in essence, that any measure of plausibility that possesses the above three properties must be a rescaling of standard probability.

While it is impressive that so much (the machinery of probability theory) can be derived from so little (Cox's very commonsensical assumptions), mathematician Michael Hardy (2002) has expressed the opinion that in fact Cox's axioms are too strong, and has provided significantly weaker conditions that lead to the same end result as Cox's three properties. Hardy's conditions are more abstract and difficult to state without introducing a lot of mathematical mechanism, but essentially he studies mappings from propositions into ordered "plausibility" values, and he shows that if any such mapping obeys the properties of

1. If x implies y then $f(x) < f(y)$
2. If $f(x) < f(y)$ then $f(\neg x) > f(\neg y)$, where $\neg$ represents "not"
3. If $f(x|z) <= f(y|z)$ and $f(x|\neg z) <= f(y|\neg z)$ then $f(x) < f(y)$
4. For all x, y either $f(x) \leq f(y)$ or $f(y) \leq f(x)$

then it maps propositions into scaled probability values. Note that this property list mixes up absolute probabilities f() with conditional probabilities f(|), but this is not a problem because Hardy considers f(x) as equivalent to f(x|U) where U is the assumed universe of discourse.

Hardy expresses regret that his fourth property is required; however, Youssef's (1994) work related to Cox's axioms suggests that it is probably there in his mathematics for a very good conceptual reason. Youssef has shown that it is feasible to drop Cox's assumption that uncertainty must be quantified using real numbers, but retain Cox's other assumptions. He shows it is possible, consistent with Cox's other assumptions, to quantify uncertainty using "numbers" drawn from the complex, quaternion, or octonion number fields. Further, he argues that complex-valued "probabilities" are the right way to model quantum-level phenomena that have not been collapsed (decohered) into classical phenomena. We believe his line of argument is correct and quite possibly profound, yet it does not seem to cast doubt on the position of standard real-valued probability theory as the correct mathematics for reasoning about ordinary, decohered physical systems. If one wishes to reason about the uncertainty existing in pure, pre-decoherence quantum systems or other exotic states of being, then arguably these probability theories defined over different base fields than the real numbers may be applicable.

Next, while we are avid probabilists, we must distinguish ourselves from the most ardent advocates of the "Bayesian" approach to probabilistic inference. We understand the weakness of the traditional approach to statistics with its reliance on often unmotivated assumptions regarding the functional forms of probability distributions. On the other hand, we don't feel that the solution is to replace these assumptions with other, often unmotivated assumptions about prior probability distributions. Bayes' rule is an important part of probability theory, but the way that the Bayesian-statistical approach applies it is not always the most useful way. A major example of the shortcomings of the standard Bayesian approach lies in the domain of confidence assessment, an important aspect of PLN already mentioned above. As Wang (2001) has argued in detail, the standard Bayesian approach does not offer any generally viable way to assess or reason about the "second-order uncertainty" involved in a given uncertainty value. Walley (1991) sought to redress this problem via a subtler approach that avoids assuming a single prior distribution, and makes a weaker assumption involving drawing a prior from a parametrized family of possible prior distributions; others have followed up his work in interesting ways (Weichselberger 2003), but this line of research has not yet been developed to the point of yielding robustly applicable mathematics. Within PLN, we introduce a spectrum of approaches to confidence assessment ranging from indefinite probabilities (essentially a hybridization of Walley's imprecise probabilities with Bayesian credible intervals) to frankly non-probabilistic heuristics inspired partly by Wang's work. By utilizing this wide range of approaches, PLN can more gracefully assess confidence in diverse settings, providing pragmatic solutions where the Walley-type approach (in spite of its purer probabilism) currently fails.

Though Cox's theorem and related results argue convincingly that probability theory is the correct approach to reasoning under uncertainty, the particular ways of applying probability theory that have emerged in the contemporary scientific community (such as the "Bayesian approach") all rely on specific assumptions beyond those embodied in the axioms of probability theory. Some of these assumptions are explicit mathematical ones, and others are implicit assumptions about how to proceed in setting up a given problem in probabilistic terms; for instance, how to translate an intuitive understanding and/or a collection of quantitative data into mathematical probabilistic form.

1.2 PLN in the Context of Traditional Approaches to Probabilistic Logic

So, supposing one buys the notion that logic, adequately broadly construed, is essential (perhaps even central) to cognition; that appropriate integration of uncertainty into logic is an important aspect of construing logic in an adequately broad way; and also that probability theory is the correct foundation for treatment of uncertainty, what then? There is already a fairly well fleshed-out theory of probabilistic logic, so why does one need a substantial body of new theory such as Probabilistic Logic Networks?

The problem is that the traditional theories in the area of probabilistic logic don't directly provide a set of tools one can use to structure a broadly-applicable, powerful software system for probabilistic inferencing. They provide a number of interesting and important theorems and ideas, but are not sufficiently pragmatic in orientation, and also fail to cover some cognitively key aspects of uncertain inference such as intensional inference.

Halpern's (2003) book provides a clearly written, reasonably thorough overview of recent theories in probabilistic logic. The early chapters of Hailperin (1996) gives some complementary historical and theoretical background. Alongside other approaches such as possibility theory, Halpern gives an excellent summary of what in PLN terms would be called "first-order extensional probabilistic logic" – the interpretation and manipulation of simple logic formulas involving absolute and conditional probabilities among sets. Shortcomings of this work from a pragmatic AI perspective include:

- No guidance is provided as to which heuristic independence assumptions are most cognitively natural to introduce in order to deal with (the usual) situations where adequate data regarding dependencies is unavailable. Rather, exact probabilistic logic formulas are introduced, into which one can, if one wishes, articulate independence assumptions and then derive their consequences.
- Adequate methods for handling "second order uncertainty" are not presented, but this is critical for dealing with real-world inference situations where available data is incomplete and/or erroneous. Hailperin (1996) deals with this by looking at interval probabilities, but this turns out to rarely be useful in practice because the intervals corresponding to inference results are generally far too wide. Walley's (1991) imprecise probabilities are more powerful but have a similar weakness, and we will discuss them in more detail in Chapter 4; they also have not been integrated into any sort of powerful, general, probabilistic logic framework, though integrating them into PLN if one wished to do so would not be problematic, as will become clear.
- Various sorts of truth-values are considered, including single values, intervals, and whole probability distributions, but the problem of finding the right way to summarize a probability distribution for logical inference without utilizing too much memory or sacrificing too much information has not been adequately resolved (and this is what we have tried to resolve with the "indefinite probabilities" utilized in PLN).
- The general probabilistic handling of intensional, temporal, and causal inference is not addressed. Of course, these topics are handled in various specialized theories; e.g., Pearl's causal networks (2000), but there is no general theory of probabilistic intensional, temporal, or causal logic; yet the majority of commonsense logical inference involves these types of reasoning.
- The existing approaches to intermixing probabilistic truth-values with existential and universal quantification are conceptually flawed and often do not yield pragmatically useful results.

All in all, in terms of Halpern's general formalism for what we call first-order extensional logic, what PLN constitutes is

- A specific compacted representation of sets of probability distributions (the indefinite truth-value)
- A specific way of deploying heuristic independence assumptions; e.g., within the PLN deduction and revision rules

- A way of counting the amount of evidence used in an inference (which is used in the revision rule, which itself uses amount of evidence together with heuristic independence assumptions)

But much of the value of PLN lies in the ease with which it extends beyond first-order extensional logic. Due to the nature of the conceptual and mathematical formalism involved, the same essential inference rules and formulas used for first-order extensional logic are extended far more broadly, to deal with intensional, temporal, and causal logic, and to deal with abstract higher-order inference involving complex predicates, higher-order functions, and universal, existential, and fuzzy quantifiers.

1.2.1 Why Term Logic?

One of the major ways in which PLN differs from traditional approaches to probabilistic logic (and one of the secrets of PLN's power) is its reliance on a formulation of logic called "term logic." The use of term logic is essential, for instance, to PLN's introduction of cognitively natural independence assumptions and to PLN's easy extension of first-order extensional inference rules to more general and abstract domains.

Predicate logic and term logic are two different but related forms of logic, each of which can be used both for crisp and uncertain logic. Predicate logic is the most familiar kind, where the basic entity under consideration is the "predicate," a function that maps argument variables into Boolean truth-values. The argument variables are quantified universally or existentially.

On the other hand, in term logic, which dates back at least to Aristotle and his notion of the syllogism, the basic element is a subject-Predicate statement, denotable

```
A → B
```

where → denotes a notion of inheritance or specialization. Logical inferences take the form of "syllogistic rules," which give patterns for combining statements with matching terms. (We don't use the → notation much in PLN, because it's not sufficiently precise for PLN purposes, since PLN introduces many varieties of inheritance; but we will use the → notation in this section because here we are speaking about inheritance in term logic in general rather than about PLN in particular.)

Examples are the deduction, induction, and abduction rules:

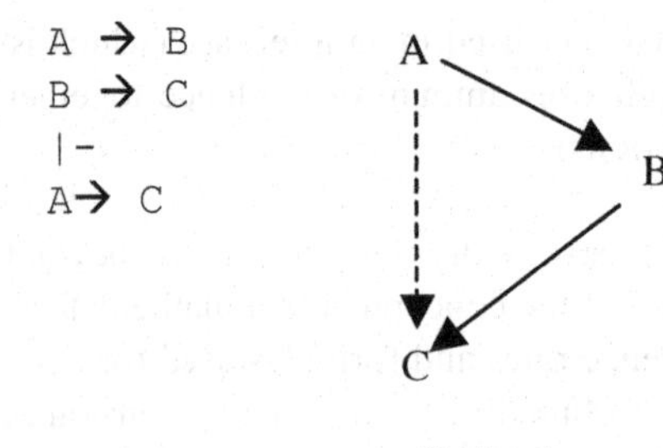

Deduction

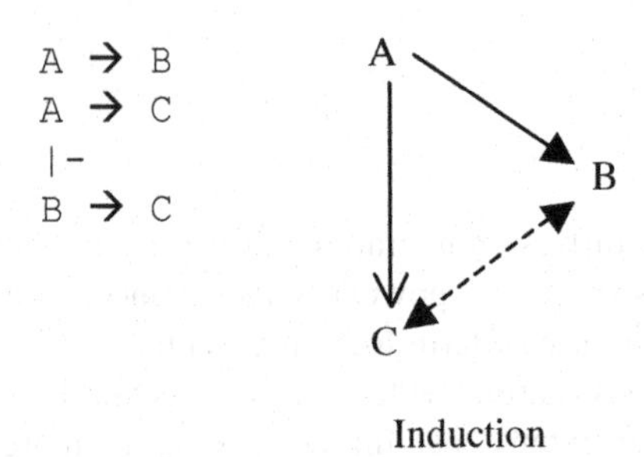

Induction

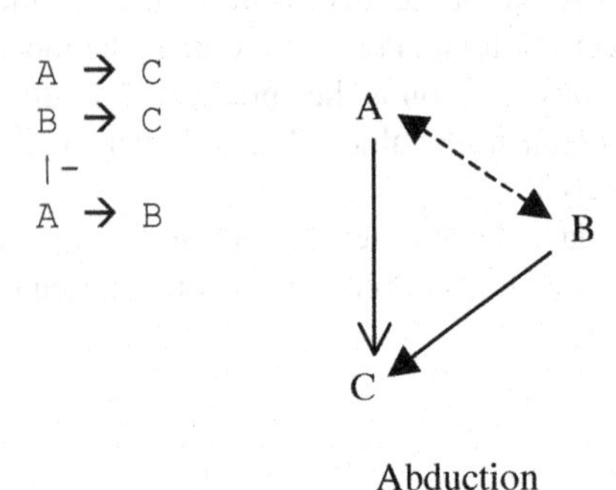

Abduction

When we get to defining the truth-value formulas corresponding to these inference rules, we will observe that deduction is infallible in the case of absolutely certain premises, but uncertain in the case of probabilistic premises; while abduction and induction are always fallible, even given certain premises. In fact we will derive abduction and induction from the combination of deduction with a simple rule called inversion

```
A → B
|-
B → A
```

whose truth-value formula derives from Bayes rule.

Predicate logic has proved to deal more easily with deduction than with induction, abduction, and other uncertain, fallible inference rules. On the other hand, term logic can deal quite elegantly and simply with all forms of inference. Furthermore, the predicate logic formulation of deduction proves less amenable to "probabilization" than the term logic formulation. It is for these reasons, among others, that the foundation of PLN is drawn from term logic rather than from predicate logic. PLN begins with a term logic foundation, and then adds on elements of probabilistic and combinatory logic, as well as some aspects of predicate logic, to form a complete inference system, tailored for easy integration with software components embodying other (not explicitly logical) aspects of intelligence.

Sommers and Engelbretsen (2000) have given an excellent defense of the value of term logic for crisp logical inference, demonstrating that many pragmatic inferences are far simpler in term logic formalism than they are in predicate logic formalism. On the other hand, the pioneer in the domain of uncertain term logic is Pei Wang (Wang 1996), to whose NARS uncertain term logic based reasoning system PLN owes a considerable debt. To frame the issue in terms of our above discussion of PLN's relation to traditional probabilistic logic approaches, we may say we have found that the formulation of appropriate heuristics to guide probabilistic inference in cases where adequate dependency information is not available, and appropriate methods to extend first-order extensional inference rules and formulas to handle other sorts of inference, are both significantly easier in a term logic rather than predicate logic context. In these respects, the use of term logic in PLN is roughly a probabilization of the use of term logic in NARS; but of course, there are many deep conceptual and mathematical differences between PLN and NARS, so that the correspondence between the two theories in the end is more historical and theory-structural, rather than a precise correspondence on the level of content.

1.3 PLN Knowledge Representation and Inference Rules

In the next few sections of this Introduction, we review the main topics covered in the book, giving an assemblage of hints as to the material to come. First, Chapter 2 describes the knowledge representation underlying PLN, without yet saying anything specific about the management of numbers quantifying uncertainties. A few tricky issues occur here, meriting conceptual discussion. Even though PLN knowledge representation is not to do with uncertain inference per se, we have found that without getting the knowledge representation right, it is very difficult to define uncertain inference rules in an intuitive way. The biggest influence on PLN's knowledge representation has been Wang's NARS framework, but there are also some significant deviations from Wang's approach.

PLN knowledge representation is conveniently understood according to two dichotomies: extensional vs. intensional, and first-order vs. higher-order. The former is a conceptual (philosophical/cognitive) distinction between logical relationships that treat concepts according to their members versus those that treat concepts according to their properties. In PLN extensional knowledge is treated as more basic, and intensional knowledge is defined in terms of extensional knowledge via the addition of a specific mathematics of intension (somewhat related to information theory). This is different from the standard probabilistic approach, which contains no specific methods for handling intension and also differs from, e.g., Wang's approach in which intension and extension are treated as completely symmetric, with neither of them being more basic or derivable from the other.

The first-order versus higher-order distinction, on the other hand, is essentially a mathematical one. First-order, extensional PLN is a variant of standard term logic, as originally introduced by Aristotle in his Logic and recently elaborated by theorists such as Wang (1996) and Sommers and Engelbretsen (2000). First-order PLN involves logical relationships between terms representing concepts, such as

```
Inheritance cat animal

ExtensionalInheritance Pixel_444 Contour_7565
```

(where the notation is used that R A B denotes a logical relationship of type R between arguments A and B). A typical first-order PLN inference rule is the standard term-logic deduction rule

```
A → B
B → C
|-
A → C
```

which in PLN looks like

```
ExtensionalInheritance A B
ExtensionalInheritance B C
|-
ExtensionalInheritance A C
```

As well as purely logical relationships, first-order PLN also includes a fuzzy set membership relationship, and specifically addresses the relationship between fuzzy set membership and logical inheritance, which is closely tied to the PLN concept of intension. In the following text we will sometimes use the acronym FOI to refer to PLN First Order Inference.

Higher-order PLN, on the other hand (sometimes called HOI, for Higher Order Inference), has to do with functions and their arguments. Much of higher-order PLN is structurally parallel to first-order PLN; for instance, implication between statements is largely parallel to inheritance between terms. However, a key difference is that most of higher-order PLN involves either variables or higher-order functions (functions taking functions as their arguments). So for instance one might have

```
ExtensionalImplication
    Inheritance $X cat
    Evaluation eat ($X, mice)
```

(using the notation that

```
R
    A
    B
```

denotes the logical relationship R applied to the arguments A and B). Here Evaluation is a relationship that holds between a predicate and its argument-list, so that, e.g.,

```
Evaluation eat (Sylvester, mice)
```

means that the list *(Sylvester, mice)* lies within the set of ordered pairs characterizing the *eat* relationship. The parallel of the first-order extensional deduction rule given above would be a rule

```
ExtensionalImplication A B
ExtensionalImplication B C
|-
ExtensionalImplication A C
```

where the difference is that in the higher-order inference case the tokens A, B, and C denote either variable-bearing expressions or higher-order functions. Some higher-order inference rules involve universal or existential quantifiers as well.

While first-order PLN adheres closely to the term logic framework, higher-order PLN is better described as a mix of term logic, predicate logic, and combinatory logic. The knowledge representation is kept flexible, as this seems to lead to the simplest and most straightforward set of inference rules.

1.4 Truth-value Formulas

We have cited above the conceptual reasons why we have made PLN a probabilistic inference framework, rather than using one of the other approaches to uncertainty quantification available in the literature. However, though we believe in the value of probabilities we do not believe that the conventional way of using probabilities to represent the truth-values of propositions is adequate for pragmatic computational purposes. One of the less conventional aspects of PLN is the quantification of uncertainty using truth-values that contain at least two components, and usually more (in distinction from the typical truth-value used in probability theory, which is a single number: a probability). Our approach here is related to earlier multi-component truth-value approaches due to Keynes (2004), Wang (2006), Walley (1991), and others, but is unique in its particulars.

The simplest kind of PLN truth-value, called a SimpleTruthValue, consists of a pair of numbers <*s*,*w*> called a strength and a confidence. The strength value is a probability; the confidence value is a measure of the amount of certainty attached to the strength value. Confidence values are normalized into [0,1].

For instance <.6,1> means a probability of .6 known with absolute certainty. <.6,.2> means a probability of .6 known with a very low degree of certainty. <.6,0> means a probability of .6 known with a zero degree of certainty, which indicates a meaningless strength value, and is equivalent to <*x*,0> for any other probability value x.

Another type of truth-value, more commonly used as the default within PLN, is the IndefiniteTruthValue. We introduce the mathematical and philosophical foundations of IndefiniteTruthValues in Chapter 4. Essentially a hybridization of Walley's imprecise probabilities and Bayesian credible intervals, indefinite probabilities quantify truth-values in terms of four numbers <*L*, *U*, *b*, *k*>: an interval $[L,U]$, a credibility level b, and an integer k called the "lookahead." IndefiniteTruthValues provide a natural and general method for calculating the "weight-of-evidence" underlying the conclusions of uncertain inferences. We ardently believe that this approach to uncertainty quantification may be adequate to serve as an ingredient of powerful artificial general intelligence.

Beyond the SimpleTruthValues and IndefiniteTruthValues mentioned above, more advanced types of PLN truth-value also exist, principally "distributional truth-values" in which the strength value is replaced by a matrix approximation to an entire probability. Note that this, then, provides for three different granularities of approximations to an entire probability distribution. A distribution can be most simply approximated by a single number, somewhat better approximated by a probability interval, and even better approximated by an entire matrix.

Chapter 5 takes the various inference rules defined in Chapter 2, and associates a "strength value formula" with each of them (a formula determining the strength of the conclusion based on the strengths of the premises). For example, the deduction rule mentioned above is associated with two strength formulas, one based on an independence assumption and the other based on a different "concept geometry" based assumption. The independence-assumption-based deduction strength formula looks like

```
B <s_B>
C <s_C>
ExtensionalInheritance A B <s_AB>
ExtensionalInheritance B C <s_BC>
|-
ExtensionalInheritance A C <s_AC>
s_AC = s_AB s_BC + (1-s_AB) ( s_C - s_B s_BC ) / (1- s_B )
```

This particular rule is a straightforward consequence of elementary probability theory. Some of the other formulas are equally straightforward, but some are subtler and require heuristic reasoning beyond standard probabilistic tools like independence assumptions. Since simple truth-values are the simplest and least informative of our truth-value types; they provide quick, but less accurate, assessments of the resulting strength and confidence values.

We reconsider these strength formulas again in Chapter 6, extending the rules to IndefiniteTruthValues. We also illustrate in detail how indefinite truth-values provide a natural approach to measuring weight-of-evidence. IndefiniteTruthValues can be thought of as approximations to entire distributions, and so provide an intermediate level of accuracy of strength and confidence.

As shown in Chapter 7, PLN inference formulas may also be modified to handle entire distributional truth-values. Distributional truth-values provide more information than the other truth-value types. As a result, they may also be used to yield even more accurate assessments of strength and confidence.

The sensitivity to error of several inference rule formulas for various parameter values is explored in Chapter 8. There we provide a fairly detailed mathematical and graphical examination of error magnification. We also study the possibility of deterministic chaos arising from PLN inference.

We introduce higher-order inference (HOI) in Chapter 10, where we describe the basic HOI rules and strength formulas for both simple truth-values and indefinite truth-values. We consider both crisp and fuzzy quantifiers, using indefinite probabilities, in Chapter 11; treat intensional inference in Chapter 12; and inference control in Chapter 13. Finally, we tackle the topics of temporal and causal inference in Chapter 14.

1.5 Implementing and Applying PLN

The goal underlying the theoretical development of PLN has been the creation of practical software systems carrying out complex, useful inferences based on uncertain knowledge and drawing uncertain conclusions. Toward that end we have implemented most of the PLN theory described in this book as will briefly be described in Chapter 13, and used this implementation to carry out simple inference experiments involving integration with external software components such as a natural language comprehension engine and a 3D simulation world.

Chapter 14 reviews some extensions made to basic PLN in the context of these practical applications, which enable PLN to handle reasoning about temporal and causal implications. Causal inference in particular turns out to be conceptually interesting, and the approach we have conceived relates causality and intension in a satisfying way.

By far the most difficult aspect of designing a PLN implementation is inference control, which we discuss in Chapter 13. This is really a foundational conceptual issue rather than an implementational matter per se. The PLN framework just tells you what inferences can be drawn; it doesn't tell you what order to draw them in, in which contexts. Our PLN implementation utilizes the standard modalities of forward-chaining and backward-chaining inference control. However, the vivid presence of uncertainty throughout the PLN system makes these algorithms more challenging to use than in a standard crisp inference context. Put simply, the search trees expand unacceptably fast, so one is almost immediately faced with the need to use clever, experience-based heuristics to perform pruning.

The issue of inference control leads into deep cognitive science issues that we briefly mention here but do not fully explore, because that would lead too far afield from the focus of the book, which is PLN in itself. One key conceptual point that we seek to communicate, however, is that uncertain inference rules and formulas, on their own, do not compose a comprehensive approach to artificial intelligence. To achieve the latter, sophisticated inference control is also required, and controlling uncertain inference is difficult – in practice, we have found, requiring ideas that go beyond the domain of uncertain inference itself. In principle, one could take a purely probability-theoretic approach to inference control – choosing inference steps based on the ones that are most likely to yield successful conclusions based on probabilistic integration of all the available evidence. However, in practice this does not seem feasible given the current state of development of applied probability theory. Instead, in our work with PLN so far, we have taken a heuristic and integrative approach, using other non-explicitly-probabilistic algorithms to help prune the search trees implicit in PLN inference control.

As for applications, we have applied PLN to the output of a natural language processing subsystem, using it to combine premises extracted from different biomedical research abstracts to form conclusions embodying medical knowledge not contained in any of the component abstracts. We have also used PLN to learn rules controlling the behavior of a humanoid agent in a 3D simulation world; for instance, PLN learns to play "fetch" based on simple reinforcement learning stimuli.

Our current research involves extending PLN's performance in both these areas, and bringing the two areas together by using PLN to help the Novamente Cognition Engine carry out complex simulation-world tasks involving a combination of physical activity and linguistic communication. Quite probably this ongoing research will involve various improvements to be made to the PLN framework itself. Our goal in articulating PLN has not been to present an ultimate

itself. Our goal in articulating PLN has not been to present an ultimate and final approach to uncertain inference, but rather to present a workable approach that is suitable for carrying out uncertain inference comprehensively and reasonably well in practical contexts. As probability theory and allied branches of mathematics develop, and as more experience is gained applying PLN in practical contexts, we expect the theory to evolve and improve.

1.6 Relationship of PLN to Other Approaches to Uncertain Inference

Finally, having sketched the broad contours of PLN theory and related it to more traditional approaches to probabilistic logic, we now briefly discuss the relationship between PLN and other approaches to logical inference. First, the debt of PLN to various standard frameworks for crisp logical inference is clear. PLN's knowledge representation, as will be made clear in Chapter 2, is an opportunistically assembled amalgam of formalisms chosen from term logic, predicate logic and combinatory logic. Rather than seeking a pure and minimal formalism, we have thought more like programming language designers and sought a logical formalism that allows maximally compact and comprehensible representation of a wide variety of useful logical structures.

Regarding uncertainty, as noted above, as well as explicit approaches to the problem of unifying probability and logic the scientific literature contains a number of other relevant ideas, including different ways to quantify uncertainty and to manipulate uncertainty once quantified. There are non-probabilistic methods like fuzzy logic, possibility theory, and NARS. And there is a variety of probabilistic approaches to knowledge representation and reasoning that fall short of being full-on "probabilistic logics," including the currently popular Bayes nets, which will be discussed in more depth below, and Walley's theory of imprecise probabilities (Walley 1991), which has led to a significant literature (ISIPTA 2001, 2003, 2005, 2007), and has had a significant inspirational value in the design of PLN's approach to confidence estimation, as will be reviewed in detail in Chapters 4, 6, and 10.

Overall, regarding the representation of uncertainty, PLN owes the most to Pei Wang's NARS approach and Walley's theory of imprecise probabilities. Fuzzy set theory ideas are also utilized in the specific context of the PLN Member relationship. However, we have not found any of these prior approaches to uncertainty quantification to be fully adequate, and so the PLN approach draws from them ample inspiration but not very many mathematical details.

We now review the relationship of PLN to a few specific approaches to uncertainty quantification and probabilistic inference in a little more detail. In all cases the comments given here are high-level and preliminary, and the ideas discussed will be much clearer to the reader after they have read the later chapters of this book and understand PLN more fully.

1.6.1 PLN and Fuzzy Set Theory

Fuzzy set theory has proved a pragmatically useful approach to quantifying many kinds of relationships (Zadeh 1965, 1978), but we believe that its utility is fundamentally limited. Ultimately, we suggest, the fuzzy set membership degree is not a way of quantifying uncertainty – it is quantifying something else: it is quantifying partial membership. Fuzzy set membership is used in PLN as the semantics of the truth-values of special logical relationship types called Member relationships. These fuzzy Member relationships may be used within PLN inference, but they are not considered the same as logical relationships such as Inheritance or Similarity relationships whose truth-values quantify degrees of uncertainty.

Some (though nowhere near all) of the fuzzy set literature appears to us to be semantically confused regarding the difference between uncertainty and partial membership. In PLN we clearly distinguish between

- Jim belongs to degree .6 to the fuzzy set of tall people. (MemberLink semantics)
- Jim shares .6 of the properties shared by people belonging to the set of tall people (where the different properties may be weighted). (IntensionalInheritanceLink semantics)
- Jim has a .6 chance of being judged as belonging to the set of tall people, once more information about Jim is obtained (where this may be weighted as to the degree of membership that is expected to be estimated once the additional information is obtained). (IntensionalInheritanceLink, aka Subset Link, semantics)
- Jim has an overall .6 amount of tallness, defined as a weighted average of extensional and intensional information. (Inheritance Link semantics)

We suggest that the fuzzy, MemberLink semantics is not that often useful, but do recognize there are cases where it is valuable; e.g., if one wishes to declare that a stepparent and stepchild are family members with fuzzy degree .8 rather than 1.

In terms of the above discussion of the foundations of probability theory we note that partial membership assignments need not obey Cox's axioms and need not be probabilities – which is fine, as they are doing something different, but also limits the facility with which they can be manipulated. In PLN, intensional probabilities are used for many of the purposes commonly associated with fuzzy membership values, and this has the advantage of keeping more things within a probabilistic framework.

1.6.2 PLN and NARS

Pei Wang's NARS approach has already been discussed above and will pop up again here and there throughout the text; furthermore, Appendix A1 presents a comparison of some of the first-order PLN truth-value formulas with corresponding NARS formulas. As already noted, there is a long historical relationship between PLN and NARS; PLN began as part of a collaboration with NARS's creator Pei Wang as an attempt to create a probabilistic analogue to NARS. PLN long ago diverged from its roots in NARS and has grown in a very different direction, but there remain many similarities. Beneath all the detailed similarities and differences, however, there is a deep and significant difference between the two, which is semantic: PLN's semantics is probabilistic, whereas NARS's semantics is intentionally and definitively not.

PLN and NARS have a similar division into first-order versus higher-order inference, and have first-order components that are strictly based on term logic. However, PLN's higher-order inference introduces predicate and combinatory logic ideas, whereas NARS's higher-order inference is also purely term logic based. Both PLN and NARS include induction, deduction, and abduction in their first-order components, with identical graphical structures; in PLN, however, induction and abduction are derived from deduction via Bayes rule, whereas in NARS they have their own completely independent truth-value functions. Both PLN and NARS utilize multi-component truth-values, but the semantics of each component is subtly different, as will be reviewed in appropriate points in the text to follow.

1.6.3 PLN and Bayes Nets

Bayes nets are perhaps the most popular contemporary approach to uncertain inference. Because of this, we here offer a few more detailed comments on the general relationship between PLN and Bayes nets. Of course, the actual relationship is somewhat subtle and will be clear to the reader only after completing the exposition of PLN.

Traditional Bayesian nets assume a tree structure for events, which is unrealistic in general, but in recent years there has been a batch of work on "loopy Bayesian networks" in which standard Bayesian net information propagation is applied to potentially cyclic graphs of conditional probability. Some interesting alternatives to the loopy Bayesian approach have also been proposed, including one that uses a more advanced optimization algorithm within the Bayesian net framework.

Bayes nets don't really contain anything comparable to the generality of PLN higher-order inference. However, in the grand scheme of things, first-order PLN is not all that tremendously different from loopy Bayesian nets and related schemes. In both cases one is dealing with graphs whose relationships denote conditional probabilities, and in both cases one is using a kind of iterative relaxation method to arrive at a meaningful overall network state.

If one took a forest of loopy Bayes nets with imprecise probabilities, and then added some formalism to interface it with fuzzy, predicate, and combinatory logic, then one might wind up with something reasonably similar to PLN. We have not taken such an approach but have rather followed the path that seemed to us more natural, which was to explicitly shape a probabilistic inference framework based on the requirements that we found important for our work on integrative AI.

There are many ways of embodying probability theory in a set of data structures and algorithms. Bayes nets are just one approach. PLN is another approach and has been designed for a different purpose: to allow basic probabilistic inference to interact with other kinds of inference such as intensional inference, fuzzy inference, and higher-order inference using quantifiers, variables, and combinators. We have found that for the purpose of interfacing basic probabilistic inference with these other sorts of inference, the PLN approach is a lot more convenient than Bayes nets or other more conventional approaches.

Another key conceptual difference has to do with a PLN parameter called the "context." In terms of probability theory, one can think of a context as a universe of discourse. Rather than attempting to determine a (possibly non-existent) universal probability distribution that has desired properties within each local domain, PLN creates local probability distributions based on local contexts. The context parameter can be set to Universal (everything the system has ever seen), Local (only the information directly involved in a given inference), or many levels in between.

Yet another major conceptual difference is that PLN handles multivariable truth-values. Its minimal truth-value object has two components: strength and weight of evidence. Alternatively, it can use probability distributions (or discrete approximations thereof) as truth-values. This makes a large difference in the handling of various realistic inference situations. For instance, the treatment of "weight of evidence" in PLN is not a purely mathematical issue, but reflects a basic conceptual issue, which is that (unlike most probabilistic methods) PLN does not assume that all probabilities are estimated from the same sample space. It makes this assumption provisionally in some cases, but it doesn't make it axiomatically and comprehensively.

With the context set to Universal, and with attention restricted to the strength component of truth-values, what we have in PLN-FOI is – speaking conceptually rather than mathematically – a different way of doing the same thing that loopy Bayes networks (BN) and its competitors are trying to do. PLN, loopy BN, and other related methods are all viewable as optimization algorithms trying to relax into a condition giving the "correct probability distribution," and at some risk of settling into local optima instead. But the ability to use more flexible truth-values, and to use local contexts as appropriate, makes a very substantial difference in practice. This is the kind of difference that becomes extremely apparent when one seeks to integrate probabilistic inference with other cognitive processes. And it's the kind of difference that is important when trying to extend one's reasoning system from simple inferences to extremely general higher-order inference – an extension that has succeeded within PLN, but has not been successfully carried out within these other frameworks.

1.7 Toward Pragmatic Probabilistic Inference

Perhaps the best way to sum up the differences between PLN and prior approaches to (crisp or uncertain) inference is to refer back to the list of requirements given toward the start of this Introduction. These requirements are basically oriented toward the need for an approach to uncertain inference that is adequate to serve as the core of a general-purpose cognition process – an approach that can handle *any kind of inference* effectively, efficiently, and in an uncertainty-savvy way.

Existing approaches to crisp inference are not satisfactory for the purposes of general, pragmatic, real-world cognition, because they don't handle uncertainty efficiently and gracefully. Of course, one can represent uncertainties in predicate logic – one can represent *anything* in predicate logic – but representing them in a way that leads to usefully rapid and convenient inference incorporating uncertainties intelligently is another matter.

On the other hand, prior approaches to uncertain inference have universally failed the test of comprehensiveness. Some approaches, such as Bayes nets and fuzzy set theory, are good at what they do but carry out only very limited functions compared to what is necessary to fulfill the inference needs of a general-purpose cognitive engine. Others, such as imprecise probability theory, are elegant and rigorous but are so complex that the mathematics needed to apply them in practical situations has not yet been resolved. Others, such as NARS and Dempster-Shafer theory, appear to us to have fundamental conceptual flaws in spite of their interesting properties. And still others, such as traditional probabilistic logic as summarized by Halpern and Hailperin, fail to provide techniques able to deal with the scale, incompleteness, and erroneousness typifying real-world inference situations.

In sum, we do not propose PLN as an ultimate and perfect uncertain inference framework, only as an adequate one – but we do suggest that, in its adequacy, PLN distinguishes itself from the alternatives currently available. As noted above, we suspect that the particulars of the PLN framework will evolve considerably as PLN is utilized for more and more pragmatic inference tasks, both on its own and within integrative AI systems.

Chapter 2: Knowledge Representation

Abstract In chapter 2, we review the basic formalism of PLN knowledge representation in a way that is relatively independent of the particularities of PLN truth-value manipulation. Much of this material has nothing explicitly to do with probability theory or uncertainty management; it merely describes a set of conventions for representing logical knowledge. However, we also define some of the elements of PLN truth-value calculation here, insofar as is necessary to define the essential meanings of some of the basic PLN constructs.

2.1 Basic Terminology and Notation

The basic players in PLN knowledge representation are entities called *terms* and *relationships* (atomic formulae). The term *Atom* will refer to any element of the set containing both terms and relationships. The hierarchy of PLN Atoms begins with a finite set *S* of elementary terms. (In an AI context, these may be taken as referring to atomic perceptions or actions, and mathematical structures.) The set of ordered and unordered subsets of *S* is then constructed, and its elements are also considered as terms. Relationships are then defined as tuples of terms, and higher-order relationships are defined as predicates or functions acting on terms or relationships.

Atoms are associated with various data items, including

- Labels indicating type; e.g., a term may be a Concept term or a Number term; a relationship may be an Inheritance relationship or a Member relationship
- Packages of numbers representing "truth-value" (more on that later)
- In some cases, Atom-type-specific data (e.g., Number terms are associated with numbers; Word terms are associated with character strings)

We will sometimes refer to uncertain truth-values here in a completely abstract way, via notation such as <*t*>. However, we will also use some specific truth-value types in a concrete way:

- "strength" truth-values, which consist of single numbers; e.g., <*s*> or <.8>. Usually strength values denote probabilities but this is not always the case. The letter *s* will be habitually used to denote strength values.
- SimpleTruthValues, which consist of pairs of numbers. These pairs come in two forms:
    - the default, <*s*,*w*>, where s is a strength and *w* is a "weight of evidence" – the latter being a number in [0,1] that tells you,

qualitatively, how much you should believe the strength estimate. The letter *w* will habitually be used to denote weight of evidence values.

  - <s,N>, where N is a "count" – a positive number telling you, qualitatively, the total amount of evidence that was evaluated in order to assess s. There is a heuristic formula interrelating w and N, w=N/(N+k) where k is an adjustable parameter. The letter N will habitually be used to denote count. If the count version rather than the weight of evidence version is being used, this will be explicitly indicated, as the former version is the default.

- IndefiniteTruthValues, which quantify truth-values in terms of four numbers <[*L*,*U*],*b*,*k*>, an interval [*L*,*U*], a credibility level *b*, and an integer *k* called the lookahead. While the semantics of IndefiniteTruthValues are fairly complex, roughly speaking they quantify the idea that after *k* more observations there is a probability *b* that the conclusion of the inference will appear to lie in the final interval [*L*,*U*]. The value of the integer *k* will often be considered a system-wide constant. In this case, IndefiniteTruthValues will be characterized more simply via the three numbers <[*L*,*U*], *b*>.

- DistributionalTruthValues, which are discretized approximations to entire probability distributions. When using DistributionalTruthValues, PLN deduction reduces simply to matrix multiplication, and PLN inversion reduces to matrix inversion.[1]

The semantics of these truth-values will be reviewed in more depth in later chapters, but the basic gist may be intuitable from the above brief comments.

PLN inference rules are associated with particular types of terms and relationships; for example, the deduction rule mentioned in the Introduction is associated with ExtensionalInheritance and Inheritance relationships. At the highest level we may divide the set of PLN relationships into the following categories, each of which corresponds to a set of different particular relationship types:

- Fuzzy membership (the Member relationship)
- First-order logical relationships
- Higher-order logical relationships
- Containers (lists and sets)
- Function execution (the ExecutionOutput relationship)

To denote a relationship of type R, between Atoms *A* and *B*, with truth-value *t*, we write

```
R A B <t>
```

If *A* and *B* have long names, we may use the alternate notation

[1] We have so far developed two flavors of DistributionalTruthValues, namely StepFunctionTruthValues and PolynomialTruthValues.

```
R <t>
    A
    B
```

which lends itself to visually comprehensible nesting; e.g.,

```
R <t>
    A
    R1
          C
          D
```

Similarly, to denote a term *A* with truth-value *t*, we write

```
    A <t>
```

For example, to say that *A* inherits from *B* with probability .8, we write

```
Inheritance A B <.8>
```

To say that *A* inherits from *B* with IndefiniteTruthValue represented by <[.8,.9], .95>, we write

```
Inheritance A B <[.8,.9],.95>
```

(roughly, as noted above, the [.8, .9] interval represents an interval probability and the .95 represents a credibility level).

We will also sometimes use object-field notation for truth-value elements, obtaining, for example, the strength value object associated with an Atom

```
(Inheritance A B).strength = [.8,.9]
```

or the entire truth-value, using .tv

```
(Inheritance A B).tv = <[.8,.9], .9, 20>.
```

Finally, we will sometimes use a semi-natural-language notation, which will be introduced a little later on, when we first get into constructs of sufficient complexity to require such a notation.

2.2 Context

PLN TruthValues are defined relative to a Context. The default Context is the entire universe, but this is not usually a very useful Context to consider. For instance, many terms may be thought of as denoting categories; in this case, the strength of a term in a Context denotes the probability that an arbitrary entity in the Context is a member of the category denoted by the term.

Contextual relationships are denoted by Context relationships, introduced in Chapter 10 The semantics of

```
Context
       C
       R A B <t>
```

is simply

```
R (A AND C) (B AND C) <t>
```

Most of the discussion in following chapters will be carried out without explicit discussion of the role of context, and yet due to the above equivalence the conclusions will be usable in the context of contextual inference.

2.3 Fuzzy Set Membership

As a necessary preliminary for discussing the PLN logical relationships, we now turn to the Member relationship. The relationship

```
Member A B <t>
```

spans two terms where the target *B* cannot be an atomic term or a relationship, but must be a term denoting a set (be it a set of atomic terms, a set of composite terms, a set of relationships, etc.). In essence, the Member relationship of PLN is the familiar "fuzzy set membership" (Zadeh 1989). For instance, we may say

```
Member Ben Goertzel_family <1>
Member Tochtli Goertzel_family <.5>
```

(Tochtli is the dog of a Goertzel family member.) When a Member relationship has a value between 0 and 1, as in the latter example, it is interpreted as a fuzzy value rather than a probabilistic value. PLN is compatible with many different algebras for combining fuzzy truth-values, including the standard min and max operators according to which if

```
Member Ben A <r>
Member Ben B <s>
```

then

```
Member Ben (A OR B) <max(r,s)>
Member Ben (A AND B) <min(r,s)>
```

When to use fuzzy set membership versus probabilistic inheritance is a somewhat subtle issue that will be discussed later on. For instance, the fuzzy set community is fond of constructs such as

```
Member Ben tall <.75>
```

which indicates that Ben is somewhat tall. But, while this is a correct PLN construct, it is also interesting in PLN to say

```
IntensionalInheritance Ben tall <.75>
```

which states that (roughly speaking) Ben shares .75 of the characteristic properties of tall things. This representation allows some useful inferences that the Member relationship does not; for instance, inheritance relationships are probabilistically transitive whereas Member relationships come without any comparably useful uncertain transitivity algebra. (As a parenthetical note, both of these are actually bad examples; they should really be

```
Context
   People
   Member Ben tall <.75>
Context
    People
    IntensionalInheritance Ben tall <.75>
```

because Ben's .75 tallness, however you define it, is not meaningful in comparison to the standard of the universe, but only in comparison to the standard of humans.)

We extend this notion of fuzzy set membership to other truth-value types as well. For instance, using IndefiniteTruthValues

```
MemberLink Astro Jetson_family <[.8,1],.95,2>
```

would mean that after 2 more observations of Astro the assessed fuzzy membership value for

```
MemberLink Astro Jetson_family
```

would lie within [.8,1] with confidence .95.

2.4 First-Order Logical Relationships

In this section, we begin our review of the PLN first-order logical relationship types, which are the following:

- Relationships representing first-order conditional probabilities:
    - Subset (extensional)
    - Inheritance (mixed)
    - IntensionalInheritance (intensional)
- Relationships representing symmetrized first-order conditional probabilities:
    - ExtensionalSimilarity (extensional)
    - Similarity (mixed)
    - IntensionalSimilarity (intensional)

- Relationships representing higher-order conditional probabilities:
  - ExtensionalImplication
  - Implication (mixed)
  - IntensionalImplication
- Relationships representing symmetrized higher-order conditional probabilities:
  - ExtensionalEquivalence
  - Equivalence (mixed)
  - IntensionalInheritance

The semantics of the higher-order logical relationships will be described briefly in Section 2.6 of this chapter and in more depth later on. The truth-value formulas for inference on these higher-order relationships are the same as those for the corresponding first-order relationships. PLN-HOI (higher-order inference) also involves a number of other relationships, such as Boolean operators (AND, OR and NOT), the SatisfyingSet operator, and an infinite spectrum of quantifiers spanning the range from ForAll to ThereExists.

2.4.1 The Semantics of Inheritance

We now explain in detail the semantics of the key PLN relationship type, Inheritance. Since inheritance in PLN represents the synthesis of extensional and intensional information, we will begin by considering extensional and intensional inheritance in their pure forms.

2.4.1.1 Subset Relationships

Firstly, a Subset relationship represents a probabilistic subset relationship; i.e., purely extensional inheritance. If we have

```
Subset A B <s>
```

(where s, a strength value, is a single number in [0,1]) where A and B are two terms denoting sets, this means

```
P(B|A) = s
```

or more precisely

```
P(x in B | x in A) = s
```

If "in" is defined in terms of crisp Member relationships (with strength in each case either 0 or 1) this means

```
P( Member x B <1> | Member x A <1>) = s
```

On the other hand, if "in" is defined in terms of fuzzy Member relationships then one must define s as

$$s = \frac{\sum_x f\left((\text{Member } x\ B).\text{strength}, (\text{Member } x\ A).\text{strength}\right)}{\sum_x (\text{Member } x\ A).\text{strength}}$$

where $f(x,y)$ denotes the fuzzy set intersection function. Two options in this regard are

```
f(x,y) = min(x,y)
f(x,y) = x*y
```

In our current practical work with PLN we're using the min function.

As before, we treat other truth-value types in an analogous manner. For example, we interpret

```
Subset A B <[L,U] 0.9, 20>
```

as $P(x \text{ in } B | x \text{ in } A) \in [L, U]$ with confidence 0.9 after 20 more observations, where "in" is defined in terms of either crisp or fuzzy Member relationships as above.

2.4.1.2 Intensional Inheritance

The Subset relationship is what we call an extensional relationship – it relates two sets according to their members. PLN also deals with intensional relationships – relationships that relate sets according to the patterns that are associated with them. The mathematics of intensionality will be given in a later chapter, but here we will review the conceptual fundamentals.

First, we review the general notions of intension and extension. These have been defined in various ways by various logical philosophers, but the essential concepts are simple. The distinction is very similar to that between a word's *denotation* and its *connotation.* For instance, consider the concept "bachelor." The extension of "bachelor" is typically taken to be *all and only the bachelors in the world* (a very large set). In practical terms, it means all bachelors that are known to a given reasoning system, or specifically hypothesized by that system. On the other hand, the *intension* of "bachelor" is the set of properties of "bachelor," including principally the property of being a *man,* and the property of being *unmarried.*

Some theorists would have it that the intension of "bachelor" consists solely of these two properties, which are "necessary and sufficient conditions" for bachelorhood; PLN's notion of intension is more flexible, and it may include necessary and sufficient conditions but also other properties, such as the fact that most bachelors have legs, that they frequently eat in restaurants, etc. These other properties allow us to understand how the concept of "bachelor" might be stretched in some contexts; for instance, if one read the sentence "Jane Smith was more of a bachelor than any of the men in her apartment building," one could make a lot more sense of it using the concept "bachelors'" full PLN intension than one could make using only the necessary-and-sufficient-condition intension.

The essential idea underlying PLN's treatment of intension is to associate both fish and whale with sets of properties, which are formalized as sets of patterns – $fish_{PAT}$ and $whale_{PAT}$, the sets of patterns associated with fish and whales. We then interpret

```
IntensionalInheritance whale fish <.7>
```

as

Subset $whale_{PAT}$ $fish_{PAT}$ <.7>

We then define Inheritance proper as the disjunction of intensional and extensional (subset) inheritance; i.e.,

```
Inheritance A B <tv>
```

is defined as

```
OR <tv>
     Subset A B
     IntensionalInheritance A B
```

The nature of reasoning on Inheritance and IntensionalInheritance relationships will be reviewed in Chapter 12; prior to that we will use Subset and related extensional relationships in most of our examples.

Why do we think intensional relationships are worth introducing into PLN? This is a cognitive science rather than a mathematical question. We hypothesize that most human inference is done not using subset relationships, but rather using composite Inheritance relationships.

And, consistent with this claim, we suggest that in most cases the natural language relation "is a" should be interpreted as an Inheritance relation between individuals and sets of individuals, or between sets of individuals – not as a Subset relationship. For instance,

```
"Stripedog is a cat"
```

as conventionally interpreted is a combination extensional/intensional statement, as is

```
"Cats are animals."
```

This statement means not only that examples of cats are examples of animals, but also that patterns in cats tend to be patterns in animals.

The idea that inheritance and implication in human language and cognition mix up intension and extension is not an original one – for example, it has been argued for extensively and convincingly by Pei Wang in his writings on NARS. However, embodying this conceptual insight Wang has outlined a different mathematics that we find awkward because it manages uncertainty in a non-probabilistic way. His approach seems to us to contradict the common sense embodied in Cox's axioms, and also to lead to counterintuitive results in many practical cases. On the other hand, our approach is consistent with probability theory but introduces measures of association and pattern-intensity as additional concepts, and integrates them into the overall probabilistic framework of PLN.

Philosophically, one may ask why a pattern-based approach to intensional inference makes sense. Why, in accordance with Cox's axioms, isn't straightforward probability theory enough? The problem is – to wax semi-poetic for a moment – that the universe we live in is a special place, and accurately reasoning about it requires making special assumptions that are very difficult and computationally expensive to explicitly encode into probability theory. One special aspect of our world is what Charles Peirce referred to as "the tendency to take habits" (Peirce,1931-1958): the fact that "patterns tend to spread"; i.e., if two things are somehow related to each other, the odds are that there are a bunch of other patterns relating the two things. To encode this tendency observed by Peirce in probabilistic reasoning one must calculate $P(A|B)$ in each case based on looking at the number of other conditional probabilities that are related to it via various patterns. But this is exactly what intensional inference, as defined in PLN, does. This philosophical explanation may seem somewhat abstruse – until one realizes how closely it ties in with human commonsense inference and with the notion of inheritance as utilized in natural language. Much more is said on this topic in Goertzel (2006).

2.4.1.3 Symmetric Logical Relationships

Inheritance is an asymmetric relationship; one may also define a corresponding symmetric relationship. Specifically one may conceptualize three such relationships, corresponding to Subset, IntensionalInheritance, and Inheritance:

```
ExtensionalSimilarity A B <tv>

tv.s = purely extensional estimate of
P(x ∈B & x ∈ A | x ∈ A OR x ∈B)

IntensionalSimilarity A B <tv>
tv.s = purely intensional estimate of
P(x ∈B & x ∈ A | x ∈ A OR x ∈B)

Similarity A B <tv>
tv.s = intensional/extensional estimate of
P(x ∈B & x ∈ A | x ∈ A OR x ∈B)
```

In each of these conceptual formulas, ∈ denotes respectively for each case SubSet, IntensionalInheritance and (mixed) Inheritance

Elementary probability theory allows us to create truth-value formulas for these symmetric logical relationships from the truth-value formulas for the corresponding asymmetric ones. Therefore we will not say much about these symmetric relationships in this book; yet in practical commonsense reasoning they are very common.

2.5 Term Truth-Values

We have discussed the truth-values of first-order logical relationships. Now we turn to a related topic, the truth-values of PLN terms. Compared to relationship truth-values, term truth-values are mathematically simpler but conceptually no less subtle.

Most simply, the truth-value of an entity *A* may be interpreted as the truth-value of a certain Subset relationship:

```
A <tv>
```

means

```
Subset Universe A <tv>
```

That is, the *A*.tv denotes the percentage of the "universe" that falls into category *A*.

This is simple enough mathematically, but the question is: what is this "universe"? It doesn't have to be the actual physical universe, it can actually be any set, considered as the "universal set" (in the sense of probability theory) for a collection of inferences. In effect, then, what we've called the Universe is really a kind of "implicit context." This interpretation will become clear in Section 2.2 and Chapter 10 when we explicitly discuss contextual inference.

Sometimes one specifically wants to do inference within a narrow local context. Other times, one wants to do inference relative to the universe as a whole, and it's in this case that things get tricky. In fact, this is one of the main issues that caused Pei Wang, the creator of the non-probabilistic NARS system that partially inspired PLN, to declare probability theory an unsound foundation for modeling human inference or designing computational inference systems. His NARS inference framework is not probability-theory based and hence does not require the positing of a universal set *U*.

Our attitude is not to abandon probability theory because of its U-dependence, but rather to explicitly acknowledge that probabilistic inference is context-dependent, and to acknowledge that context selection for inference is an important aspect of cognition. When a mind wants to apply probabilistic reasoning, nothing tells it a priori how to set this particular parameter (the size of the universal set, $|U|$), which makes a big difference in the results that reasoning gives. Rather, we believe, the context for an inference must generally be determined by *non-inferential cognitive processes,* aided by appropriate inference rules.

There are two extreme cases: Universal and Local context. In the Universal case, pragmatically speaking, the set U is set equal to everything the system has ever seen or heard of. (This may of course be construed in various different ways in practical AI systems.) In the Local case, the set U is set equal to the union of the premises involved in a given inference, with nothing else acknowledged at all. According to the algebra of the PLN deduction rule, as will be elaborated below, local contexts tend to support more speculative inferences, whereas in the Universal context only the best supported of inferences come out with nontrivial strength.

In some cases, the context for one probabilistic inference may be figured out by another, separate, probabilistic inference process. This can't be a universal solution to the problem, however, because it would lead to an infinite regress. Ultimately one has to bottom out the regress, either by assuming a universal set U is given a priori via "hard wiring" (perhaps a hard-wired function that sets U adaptively based on experience) or by positing that U is determined by non-inferential processes.

If one wants to choose a single all-purpose U, one has to err on the side of inclusiveness. For instance, $|U|$ can be set to the sum of the counts of all Atoms in the system. Or it can be set to a multiple of this, to account for the fact that the system cannot afford to explicitly represent all the entities it knows indirectly to exist.

It is not always optimal to choose a universal and maximal context size, however. Sometimes one wants to carry out inference that is specifically restricted to a certain context, and in that case choosing a smaller U is necessary in order to get useful results. For instance, if one is in the USA and is reasoning about the price of furniture, one may wish to reason only in the context of the USA, ignoring all information about the rest of the world.

Later on we will describe the best approach we have conceived for defining U in practice, which is based on an equation called the "optimal universe size formula." This approach assumes that one has defined a set of terms that one wants to consider as a context (e.g., the set of terms pertaining to events or entities in the USA, or *properties of* the USA). One also must assume that for some terms A, B and C in this context-set, one has information about the triple-intersections $P(A \cap B \cap C)$. Given these assumptions, a formula may be derived that yields the U-value that is optimal, in the sense of giving the minimum error for PLN deduction in that context. Note that some arbitrariness is still left here; one must somewhere obtain the context definition; e.g., decide that it's intelligent to define U relative to the United States of America, or relative to the entire system's entire experience, etc.

This formula for deriving the value of U is based on values called "count values", representing numbers of observations underlying truth value estimates, and closely related to the confidence components of truth values. This means that the challenge of defining U ultimately bottoms out in the problem of count/confidence updating. In an integrative AI architecture, for example, two sorts of processes may be used for updating the confidence components of Atoms' TruthValues. Inference can be used to modify count values as well as strength values, which covers the case where entities are inferred to exist rather than observed to exist. And in an architecture incorporating natural language processing, one can utilize "semantic mapping schemata," which translate perceived linguistic utterances into sets of Atoms, and which may explicitly update the confidence components of truth values. To take a crude example, if a sentence says "I am very sure cats are only slightly friendly", this translates into a truth value with a low strength and high confidence attached to the Atoms representing the statement "cats are friendly." An important question there is: What process learns these cognitive schemata carrying out semantic mapping? If they are learned by probabilistic inference, then they must be learned within some universal set U. The pragmatic answer we have settled on in our own work is that inference applied to schema learning basically occurs using a local context, in which the schema known to the system are assumed to be all there are. Some of these schemata learned with a local context are then used to manipulate the count variables of other Atoms, thus creating a larger context for other applications of inference within the system.

In our practical applications of PLN, we have found it is not that often that the most universal U known to the system is used. More often than not, inference involves some relatively localized context. For example, if the system is reasoning about objects on Earth it should use a U relativized to Earth's surface, rather than using the U it has inferred for the entire physical universe. P(air) is very small in the context of the whole physical universe, but much larger in the context of the Earth. Every time the inference system is invoked it must assume a certain context size $|U|$, and there are no rigid mathematical rules for doing this. Rather, this parameter-setting task is a job for cognitive schema, which are learned by a host of processes in conjunction, including inference conducted with respect to the implicit local context generically associated with schema learning.

2.6 Higher-Order Logical Relationships

The first-order logical relationships reviewed above are all relationships between basic terms. But the same sorts of probabilistic logical relationships may be seen to hold between more complex expressions involving variables (or variable-equivalent constructs like SatisfyingSets). ExtensionalImplication, for example, is a standard logical implication between predicates. In PLN-HOI we have the notion of a predicate similar to standard predicate logic as a function that maps arguments into truth-values. We have an Evaluation relationship so that, e.g., for the predicate *isMale*,

```
Evaluation isMale Ben_Goertzel <1>
Evaluation isMale Izabela_Goertzel <0>
Evaluation isMale Hermaphroditus <0.5>
```

So if we have the relationship

```
ExtensionalImplication isMale hasPenis <.99>
```

this means

```
isMale($X) implies hasPenis($X) <.99>
```

or in other words

```
ExtensionalImplication <.99>
     Evaluation isMale $X
     Evaluation hasPenis $X
```

or

```
P( Evaluation hasPenis $X | Evaluation isMale $X) =
.99
```

Note that we have introduced a new notational convention here: the names of variables that are arguments of Predicates are preceded by the $ sign.This convention will be used throughout the book.

Regarding the treatment of the non-crisp truth-values of the Evaluation relationships, the same considerations apply here as with Subset and Member relationships. Essentially we are doing fuzzy set theory here and may use the min(,) function between the Evaluation relationship strengths. As this example indicates, the semantics of higher-order PLN relationships thus basically boils down to the semantics of first-order PLN relationships. To make this observation formal we must introduce the SatisfyingSet operator.

We define the SatisfyingSet of a Predicate as follows: *the SatisfyingSet of a Predicate is the set whose members are the elements that satisfy the Predicate.* Formally, that is:

```
S = SatisfyingSet P
```

means

```
( Member $X S ).tv = ( Evaluation P $X).tv
```

In PLN, generally speaking, one must consider not only Predicates that are explicitly embodied in Predicate objects, but also Predicates defined implicitly by relationship types; e.g., predicates like

```
P($x) = Inheritance $x A
```

This means that relationships between relationships may be considered as a special case of relationships between predicates.

In any case, given an individual Predicate h we can construct SatisfyingSet(h), and we can create an average over a whole set of Predicates h,

```
B ⊂ SatisfyingSet(h) → A ⊂ SatisfyingSet(h)
```

Thus, information about $h(x)$ for various $x \in B$ and $x \in A$ and various Predicates h can be used to estimate the strengths of subset relationships between sets.

Also note that in dealing with SatisfyingSets we will often have use for the Predicate NonEmpty, which returns 1 if its argument is nonempty and 0 if its argument is the empty set. For instance,

```
Evaluation NonEmpty (SatisfyingSet (eats_bugs)) <1>
```

means that indeed, there is somebody or something out there who eats bugs.

The main point of SatisfyingSets is that we can use them to map from higher-order into first-order. A SatisfyingSet maps Evaluation relationships into Member relationships, and hence has the side effect of mapping higher-order relations into ordinary first-order relations between sets. In other words, by introducing this *one* higher-order relationship (SatisfyingSet) as a primitive we can automatically get all other higher-order relationships as consequences. So using SatisfyingSets we don't *need* to introduce special higher-order relationships into PLN at all. However, it turns out to be convenient to introduce them anyway, even though they are "just" shorthand for expressions using SatisfyingSets and first-order logical relationships.

To understand the reduction of higher-order relations to first-order relations using SatisfyingSets, let R_1 and R_2 denote two (potentially different) relationship types and let X denote an Atom-valued variable, potentially restricted to some subclass of Atoms such as a particular term or relationship type. For example, we may construct the following higher-order relationship types:

ExtensionalImplication
    R_1 A X
    R_2 B X
equals
Subset
    SatisfyingSet(R_1 A X)
    SatisfyingSet(R_2 B X)

Implication
    R_1 A X
    R_2 B X
equals
Inheritance
    SatisfyingSet(R_1 A X)
    SatisfyingSet(R_2 B X)

ExtensionalEquivalence R_1 A X

```
    R2 B X
equals
ExtensionalSimilarity
    SatisfyingSet(R1 A X)
    SatisfyingSet(R2 B X)
```

```
Equivalence
    R1 A X
    R2 B X
equals
Similarity
    SatisfyingSet(R1 A X)
    SatisfyingSet(R2 B X)
```

Higher-order purely intensional symmetric and asymmetric logical relationships are omitted in the table, but may be defined analogously.

To illustrate how these higher-order relations work, consider an example higher-order relationship, expressed in first-order logic notation as

```
∀X (Member Ben X) → (Inheritance scientists X)
```

This comes out in PLN notation as:

```
ExtensionalImplication
    Member Ben X
    Inheritance scientists X
```

("if Ben is a member of the group *X*, then *X* must contain scientists"), or equivalently

```
Subset
    SatisfyingSet (Member Ben)
    SatisfyingSet (Inheritance scientists)
```

("the set of groups *X* that satisfies the constraint 'MemberRelationship Ben *X*' is a subset of the set of groups *X* that satisfies the constraint 'Inheritance scientists *X*.'")

While the above examples have concerned single-variable relationships, the same concepts and formalism work for the multiple-variable case, via the mechanism of using a single list-valued variable to contain a list of component variables.

2.7 N-ary Logical Relationships

The next representational issue we will address here has to do with relationships that have more than one argument. We don't just want to be able to say that *cat* inherits from *animal,* we want to be able to say that cats eat mice, that flies give diseases to people, and so forth. We want to express complex *n*-ary relations, and then reason on them.

In PLN there are two ways to express an *n*-ary relation: using list relationships, and using higher-order functions. Each has its strengths and weaknesses, so the two are used in parallel. For instance, the list approach is often more natural for inferences using *n*-ary relationships between simple terms, whereas the higher-order function approach is often more natural for many aspects of inference involving complex Predicates.

2.7.1 The List/Set Approach

The List approach to representing n-ary relations is very simple. An *n*-ary relation *f* with the arguments $x_1, \ldots, x_n$ is represented as a Predicate, where

```
Evaluation f (x1, ..., xn)
```

So for instance, the relationship "Ben kicks Ken" becomes roughly

```
Evaluation kicks (Ben, Ken)
```

This doesn't take the temporal aspect of the statement into account, but we will ignore that for the moment (the issue will be taken up later on).

In some cases one has a relationship that is symmetric with respect to its arguments. One way to represent this is to use a Set object for arguments. For instance, to say "*A* fuses with *B*" we may say

```
Evaluation fuse {A, B}
```

where $\{A, B\}$ is a Set. This kind of representation is particularly useful when one is dealing with a relationship with a large number of arguments, as often occurs with the processing of perceptual data.

2.7.2 Curried Functions

Another way to represent *n*-ary functions – in the spirit of Haskell rather than LISP – is using function currying. For example, a different representation of "Ben kicks Ken" is

```
Evaluation (kick Ben) Ken
```

where in this case the interpretation is that (kick Ben) is a function that outputs a Predicate function that tells whether its argument is kicked by Ben or not. Strictly, of course, the *kick* in this example is not the same as the *kick* in the argument list example; a more correct notation would be, for instance,

```
Evaluation kick_List (Ben, Ken)
Evaluation (kick_curry Ben) Ken
```

In a practical PLN system these two functions will have to be represented by different predicates, with the equivalence relation

```
Equivalence
    Evaluation (kick_curry $x) $y
    Evaluation kick_List ($x, $y)
```

and/or one or more of the listification relations

```
kick_List = listify kick_curry
kick_curry = unlistify kick_List
```

stored in the system to allow conversion back and forth.

Another representation is then

```
Evaluation (kick_curry_2 Ken) Ben
```

which corresponds intuitively to the passive voice "Ken was kicked by Ben." We then have the conceptual equivalences

```
kick_curry = "kicks"
kick_curry_2 = "is kicked by"
```

Note that the relation between *kick_curry* and *kick_curry_2* is trivially representable using the C combinator (note that in this book we use the notational convention that combinators are underlined) by

```
kick_curry = C kick_curry_2
```

Mathematical properties of Predicates are easily expressed in this notation. For instance, to say that the Predicate fuse is symmetric we need only use the higher-order relationship

```
EquivalenceRelationship
    fuse
    C fuse
```

or we could simply say

```
Inheritance fuse symmetric
```

where the Predicate *symmetric* is defined by

```
ExtensionalEquivalence
    Evaluation symmetric $X
    Equivalence
        $X
        C $X
```

Chapter 3: Experiential Semantics

Abstract Chapter 3 is a brief chapter in which we discuss the conceptual interpretation of the terms used in PLN, according to the scheme we have deployed when utilizing PLN to carry out inferences regarding the experience of an embodied agent in a simulated world. This is what we call "experiential semantics." The PLN mathematics may also be applied using different semantic assumptions, for instance in a logical theorem-proving context. But the development of PLN has been carried out primarily in the context of experiential semantics, and that will be our focus here.

3.1 Introduction

Most of the material in this book is mathematical and formal rather than philosophical in nature. Ultimately, however, the mathematics of uncertain logic is only useful when incorporated into a practical context involving non-logical as well as logical aspects; and the integration of logic with non-logic necessarily requires the conceptual interpretation of logical entities.

The basic idea of experiential semantics is that the interpretation of PLN terms and relationships should almost always be made in terms of the observations made by a specific system in interacting with the world. (Some of these observations may, of course, be observations of the system itself.) The numerous examples given in later (and prior) sections regarding "individuals" such as people, cats and so forth, can't really be properly interpreted without attention to this fact, and in particular to the experiential semantics of individuals to be presented here. What makes PLN's experience-based semantics subtle, however, is that there are many PLN terms and relationships that don't refer directly to anything in the world the PLN reasoning system is observing. But even for the most abstract relationships and concepts expressed in PLN, the semantics must ultimately be grounded in observations.

3.2 Semantics of Observations

In experiential semantics, we are considering PLN as a reasoning system intended for usage by an embodied AI agent: one with perceptions and actions as well as cognitions. The first step toward concretizing this perspective is to define what we mean by *observations*.

While PLN is mostly about statements with probabilistic truth values, at the most basic semantic level it begins with Boolean truth-valued statements. We call these basic Boolean truth-valued statements "elementary observations." Elementary observations may be positive or negative. A positive observation is one that occurred; a negative observation is one that did not occur. Elementary observations have the property of unity: each positive elementary observation occurs *once*. This occurrence may be effectively instantaneous or it may be spread over a long period of time. But each elementary observation occurs once and is associated with one particular set of time-points.

In the experiential-semantics approach, PLN's probabilistic statements may all ultimately be interpreted as statements about sets of elementary observations. In set-theoretic language, these elementary observations are "atoms" and PLN Concept terms are sets built up from these "atoms." However we won't use that terminology much here, since in PLN the term Atom is used to refer to any PLN term or relationship. Instead, we will call these "elementary terms." In an experiential-semantics approach to PLN, these are the basic terms out of which the other PLN terms and relationships are built.

For example, in the context of an AI system with a camera eye sensor, an elementary observation might be the observation A defined by

```
A = "the color of the pixel at location (100, 105) is
    blue at time 2:13:22PM on Tuesday January 6,
2004."
```

Each elementary observation may be said to contain a certain number of bits of information; for instance, an observation of a color pixel contains more bits of information than an observation of a black-and-white pixel.

3.2.1 Inference on Elementary Observations

Having defined elementary observations, one may wish to draw implication relationships between them. For instance, if we define the elementary observations

```
B = "blue was observed by me at time 2:13:22PM on
    Tuesday January 6, 2004"

C = "blue was observed by me on Tuesday January 6,
    2004"
```

then we may observe that

```
A implies B
B implies C
A implies C
```

However, the semantics of this sort of implication is somewhat subtle. Since each elementary observation occurs only once, there is no statistical basis on which to create implications between them. To interpret implications between elementary observations, one has to look across multiple "possible universes," and observe that for instance, in any possible universe in which A holds, C also holds. This is a valid form of implication, but it's a subtle one and occurs as a later development in PLN semantics, rather than at a foundational level.

3.2.2 Inference on Sets of Elementary Observations

Next, sets of elementary observations may be formed and their unions and intersections may be found, and on this basis probabilistic logical relationships between these sets may be constructed. For instance, if

```
X = the set of all observations of dark blue
Y = the set of all observations of blue
```

then it's easy to assign values to P(X|Y) and P(Y|X) based on experience.

```
P(X|Y) = the percentage of observations of blue that
are also observations of dark blue
P(Y|X) = 1, because all observations of dark blue are
observations of blue
```

Probabilistic inference on sets of observations becomes interesting because, in real-world intelligence, each reasoning system collects far more observations than it can retain or efficiently access. Thus it may retain the fact that

```
P(dark blue | blue) = .3
```

without retaining the specific examples on which this observation is founded. The existence of "ungrounded" probabilistic relationships such as this leads to the need for probabilistic inference using methods like the PLN rules.

In this context we may introduce the notion of "intension," considered broadly as "the set of attributes possessed by a term." This is usually discussed in contrast to "extension," which is considered as the elements in the set denoted by a term. In the absence of ungrounded probabilities, inference on sets of observations can be purely extensional. However, if a reasoning system has lost information about the elements of a set of observations, but still knows other sets the set belongs to, or various conditional probabilities relating the set to other sets, then it may reason about the set using this indirect information rather than the (forgotten) members of the set – and this reasoning may be considered "intensional." This is a relatively simple case of intensionality, as compared to intensionality among individuals and sets thereof, which will be discussed below.

3.3 Semantics of Individuals

While the basic semantics of PLN is founded on observations, most of the concrete PLN examples we will give in these pages involve individual entities (people, animals, countries, and so forth) rather than directly involving observations. The focus on individuals in this text reflects the level at which linguistic discourse generally operates, and shouldn't be taken as a reflection of the level of applicability of PLN: PLN is as applicable to perception-and-action-level elementary observations as it is to abstract inferences about individuals and categories of individuals. However, reasoning about individuals is obviously a very important aspect of commonsense reasoning; and so, in this section, we give an explicit treatment of the semantics of individuals. Reasoning regarding individuals is a somewhat subtler issue than reasoning regarding observations, because the notion of an "individual" is not really a fundamental concept in mathematical logic.

Conventional approaches to formalizing commonsense inference tend to confuse things by taking individuals as logical atoms. In fact, in the human mind or the mind of any AI system with a genuine comprehension of the world, individuals are complex cognitive constructs. Observations are much more naturally taken as logical atoms, from a mathematical and philosophical and cognitive-science point of view. However, from a practical commonsense reasoning point of view, if one takes elementary observations as logical atoms, then inference regarding individuals can easily become horribly complex, because the representation of a pragmatically interesting individual in terms of elementary observations is generally extremely complex. PLN works around this problem by synthesizing individual-level and observation-level inference in a way that allows individual-level inference to occur based *implicitly* on observation-level semantics. This is a subtle point that is easy to miss when looking at practical PLN inference examples, in which individual- and observation-level semantics are freely intermixed in a consistent way. This free intermixture is only possible because the conceptual foundations of PLN have been set up in a proper way.

Let's consider an example of an individual: the orange cat named Stripedog[1] who is sitting near me (Ben Goertzel) as I write these words. What is Stripedog? In PLN terms, Stripedog is first of all a complex predicate formed from elementary observations. Given a set of elementary observations, my mind can evaluate whether this set of observations is indicative of Stripedog's presence or not. It does so by evaluating a certain predicate whose input argument is a set of elementary observations and whose output is a truth value indicating the "degree of Stripedogness" of the observation set.

At this point we may introduce the notion of the "usefulness" of an argument for a predicate, which will be important later on. If a predicate P is applied to an observation set S, and an observation O lies in S, then we say that O is important for (P, S) if removing O from S would alter the strength of the truth value of P as applied to S. Otherwise O is unimportant for (P, S). We will say that an observation set S is an *identifier* for P if one of two conditions holds:

Positive identifier: S contains no elements that are unimportant for (P, S), or
Negative identifier: P applied to S gives a value of 0

Of course, in addition to the observational model of Stripedog mentioned above, I also have an abstract model of Stripedog in my mind. According to this abstract model, Stripedog is a certain pattern of arrangement of molecules. The individual molecules arranged in the Stripedogish pattern are constantly disappearing and being replaced, but the overall pattern of arrangement is retained. This abstract model of Stripedog exists in my mind because I have an abstract model of the everyday physical world in my mind, and I have some (largely ungrounded) implications that tell me that when a Stripedoggish elementary observation set is presented to me, this implies that a Stripedoggish pattern of arrangement of molecules is existent in the physical world that I hypothesize to be around me.

The Stripedog-recognizing predicate, call it $F_{Stripedog}$, has a SatisfyingSet that we may denote simply as *stripedog*, defined by

```
ExtensionalEquivalence
    Member $X stripedog
    AND
          Evaluation F_stripedog $X
          Evaluation isIdentifier ($X, F_stripedog)
```

The predicate *isIdentifier(S,P)* returns True if and only if the set S is an identifier for the predicate P, in the sense defined above.

---

[1] In case you're curious, "Stripedog" is a colloquialism for "badger" -- a word that Ben Goertzel's son Zebulon discovered in the Redwall books by Brian Jacques, and decided was an excellent name for a cute little orange kitten-cat.

This set, *stripedog*, is the set of all observation sets that are identifiers for the individual cat named Stripedog, with a fuzzy truth value function defined by the extent to which an observation set is identified as being an observation of Stripedog.

3.3.1 Properties of Individuals

Now, there may be many predicates that imply and/or are implied by $F_{Stripedog}$ to various degrees. For instance there's a predicate F_{living_being} that says true whenever a living being is observed; clearly

```
Implication F_Stripedog F_living_being
```

holds with a strength near 1 (a strength 1 so far, based on direct observation, since Stripedog has not yet been seen dead; but a strength <1 based on inference since it's inferred that it's not impossible, though unlikely, to see him dead). And,

```
Implication F_Stripedog F_orange<.8, .99>
```

holds as well – the .8 being because when Stripedog is seen at night, he doesn't look particularly orange.

Those predicates that are probabilistically implied by the Stripedog-defining predicate are what we call *properties* of Stripedog.
Note that if

```
Implication F G
```

holds, then

```
Inheritance (SatisfyingSet F) (SatisfyingSet G)
```

holds. So properties of Stripedog correspond to observation sets that include observations of Stripedog plus other, non-Stripedoggish observations.

3.4 Experiential Semantics and Term Probabilities

Another conceptual issue that arises in PLN related to experiential semantics is the use of term probabilities. It is reasonable to doubt whether a construct such as P(cat) or P(Stripedog) makes any common sense – as opposed to conditional probabilities denoting the probabilities of these entities in some particular contexts. In fact we believe there is a strong cognitive reason for a commonsense reasoning engine to use default term probabilities, and one that ties in with experiential semantics and merits explicit articulation.

The key point is that the default probability of a term represents the probability relative to the context of the entire universe as intellectually understood by the reasoning system. This would be nebulous to define and of extremely limited utility. Rather, a term probability should represent the probability of the class denoted by the term relative to the organism's (direct and indirect) experience.

An analogy to everyday human experience may be worthwhile. Outside of formal contexts like science and mathematics, human organisms carry out commonsense reasoning within the default context of their everyday embodied life. So for instance in our everyday thinking we assume as a default that cats are more common than three-eared wombats. Though we can override this if the specific context calls for it.

So, we conjecture, the term probability of "cat" in a typical human mind is shorthand for "the term probability of cat in the default context of my everyday life" – but it is not represented anything like this; rather, the "everyday life" context is left implicit.

Formalistically, we could summarize the above discussion by saying that: The default term probability of X is the weighted average of the probability of X across all contexts C, where each context C is weighted by its importance to the organism.

In this sense, default term probabilities become more heuristic than context-specific probabilities. And they also require concepts outside PLN for their definition, relying on the embedding of PLN in some broader embodied cognition framework such as the NCE.

The conceptual reason why this kind of default node probability is useful is that doing all reasoning contextually is expensive, as there are so many contexts. So as an approximation, assuming a default experiential context is very useful. But the subtlety is that for an organism that can read, speak, listen, and so forth, the "everyday experiential context" needs to go beyond what is directly experienced with the senses.

3.5 Conclusion

In this brief chapter, beginning with elementary observations, we have built up to individuals and their properties. The semantic, conceptual notions presented here need not be invoked explicitly when reviewing the bulk of the material in this book, which concerns the mathematics of uncertain truth value estimation in PLN. However, when interpreting examples involving terms with names like "Ben" and "cat", it is important to remember that in the context of the reasoning carried out by an embodied agent, such terms are not elementary indecomposables but rather complex constructs built up in a subtle way from a large body of elementary observations.

Chapter 4: Indefinite Truth Values

Abstract In this chapter we develop a new approach to quantifying uncertainty via a hybridization of Walley's theory of imprecise probabilities and Bayesian credible intervals. This "indefinite probability" approach provides a general method for calculating the "weight-of-evidence" underlying the conclusions of uncertain inferences. Moreover, both Walley's imprecise beta-binomial model and standard Bayesian inference can be viewed mathematically as special cases of the more general indefinite probability model.

4.1 Introduction

One of the major issues with probability theory as standardly utilized involves the very quantification of the uncertainty associated with statements that serve as premises or conclusions of inference. Using a single number to quantify the uncertainty of a statement is often not sufficient, a point made very eloquently by Wang (Wang 2004), who argues in detail that the standard Bayesian approach does not offer any generally viable way to assess or reason about the "second-order uncertainty" involved in a given probability assignment. Probability theory provides richer mechanisms than this: one may assign a probability distribution to a statement, instead of a single probability value. But what if one doesn't have the data to fill in a probability distribution in detail? What is the (probabilistically) best approach to take in the case where a single number is not enough but the available data doesn't provide detailed distributional information? Current probability theory does not address this issue adequately. Yet this is a critical question if one wants to apply probability theory in a general intelligence context. In short, one needs methods of quantifying uncertainty at an intermediate level of detail between single probability numbers and fully known probability distributions. This is what we mean by the question: *What should an uncertain truth-value be, so that a general intelligence may use it for pragmatic reasoning?*

4.2 From Imprecise Probabilities to Indefinite Probabilities

Walley's (Walley 1991) theory of imprecise probabilities seeks to address this issue, via defining interval probabilities, with interpretations in terms of families of probability distributions. The idea of interval probabilities was originally introduced by Keynes (Keynes 1921, 2004), but Walley's version is more rigorous, grounded in the theory of envelopes of probability distributions. Walley's intervals, so-called "imprecise probabilities," are satisfyingly natural and consistent in the way they handle uncertain and incomplete information. However, in spite of a fair amount of attention over the years, this line of research has not yet been developed to the point of yielding robustly applicable mathematics.

Using a parametrized envelope of (beta-distribution) priors rather than assuming a single prior as would be typical in the Bayesian approach, Walley (Walley 1991, 1996) concludes that it is plausible to represent probabilities as intervals of the form $\left[\frac{m}{n+k}, \frac{m+k}{n+k}\right]$. In this formula n represents the total number of observations, m represents the number of positive observations, and k is a parameter that Walley calls s and derives as a parameter of the beta distribution. Walley calls this parameter the learning parameter, while we will refer to it as the lookahead parameter. Note that the width of the interval of probabilities is inversely related to the number of observations n, so that the more evidence one has, the narrower the interval. The parameter k determines how rapidly this narrowing occurs. An interval of this sort is what Walley calls an "imprecise probability."

Walley's approach comes along with a host of elegant mathematics including a Generalized Bayes' Theorem. However it is not the only approach to interval probabilities. For instance, one alternative is Weichselberger's (Weichselberger, 2003) axiomatic approach, which works with sets of probabilities of the form $[L, U]$ and implies that Walley's generalization of Bayes' rule is not the correct one.

One practical issue with using interval probabilities like Walley's or Weichselberger's in the context of probabilistic inference rules (such as those used in PLN) is the pessimism implicit in interval arithmetic. If one takes traditional probabilistic calculations and simplistically replaces the probabilities with intervals, then one finds that the intervals rapidly expand to [0, 1]. This fact simply reflects the fact that the intervals represent "worst case" bounds. This same problem also affects Walley's and Weichselberger's more sophisticated approaches, and other approaches in the imprecise probabilities literature. The indefinite probabilities approach presented here circumvents these practical problems by utilizing interval probabilities that have a different sort of semantics – closely related to, but not the same as, those of Walley's interval probabilities.

Indefinite probabilities, as we consider them here, are represented by quadruples of the form $<(L, U], b, k>$– thus, they contain two additional numbers beyond the $[L, U]$ interval truth values proposed by Keynes, and one number beyond the $<(L, U], k>$ formalism proposed by Walley. The semantics involved in assigning such a truth value to a statement S is, roughly, "I assign a probability of b to the hypothesis that, after I have observed k more pieces of evidence, the truth value I assign to S will lie in the interval $[L, U]$." In the practical examples presented here we will hold k constant and thus will deal with truth value triples $<(L, U], b>$.

The inclusion of the value b, which defines the credibility level according to which $[L, U]$ is a credible interval (for hypothesized future assignments of the probability of S, after observing k more pieces of evidence), is what allows our intervals to generally remain narrower than those produced by existing imprecise probability approaches. If b=1, then our approach essentially reduces to imprecise probabilities, and in pragmatic inference contexts tends to produce intervals [L, U] that approach [0, 1]. The use of b<1 allows the inferential production of narrower intervals, which are more useful in a real-world inference context.

In practice, to execute inferences using indefinite probabilities we make heuristic distributional assumptions, assuming a "second-order" distribution which has [L, U] as a (100*b)% credible interval, and then "first-order" distributions whose means are drawn from the second-order distribution. These distributions are to be viewed as heuristic approximations intended to estimate unknown probability values existing in hypothetical future situations. The utility of the indefinite probability approach may be dependent on the appropriateness of the particular distributional assumptions to the given application situation. But in practice we have found that a handful of distributional forms seem to suffice to cover commonsense inferences (beta and bimodal forms seem good enough for nearly all cases; and here we will give only examples covering the beta distribution case).

Because the semantics of indefinite probabilities is different from that of ordinary probabilities, or imprecise probabilities, or for example NARS truth values, it is not possible to say objectively that any one of these approaches is "better" than the other one, as a mathematical formalism. Each approach is better than the others at mathematically embodying its own conceptual assumptions. From an AGI perspective, the value of an approach to quantifying uncertainty lies in its usefulness when integrated with a pragmatic probabilistic reasoning engine. While complicated and dependent on many factors, this is nevertheless the sort of evaluation that we consider most meaningful.

Section 4.3 deals with the conceptual foundations of indefinite probabilities, clarifying their semantics in the context of Bayesian and frequentist philosophies of probability. Section 4.4 outlines the pragmatic computational method we use for doing probabilistic and heuristic inference using indefinite probabilities.

4.3 The Semantics of Uncertainty

The main goal of this chapter is to present indefinite probabilities as a pragmatic tool for uncertain inference, oriented toward utilization in AGI systems. Before getting practical, however, we will pause in this section to discuss the conceptual, semantic foundations of the "indefinite probability" notion. In the course of developing the indefinite probabilities approach, we found that the thorniest aspects lay not in the mathematics or software implementation, but rather in the conceptual interpretation of the truth values and their roles in inference.

In the philosophy of probability, there are two main approaches to interpreting the meaning of probability values, commonly labeled frequentist and Bayesian (Stanford Encyclopedia of Philosophy 2003). There are many shades of meaning to each interpretation, but the essential difference is easy to understand. The frequentist approach holds that a probability should be interpreted as the limit of the relative frequency of an event-category, calculated over a series of events as the length of the series tends to infinity. The subjectivist or Bayesian approach holds that a probability should be interpreted as the degree of belief in a statement, held by some observer; or in other words, as an estimate of how strongly an observer believes the evidence available to him supports the statement in question. Early proponents of the subjectivist view were Ramsey (Ramsey 1931) and de Finetti (de Finetti 1974-75), who argued that for an individual to display self-consistent betting behavior they would need to assess degrees of belief according to the laws of probability theory. More recently Cox's theorem (Cox 1946) and related mathematics (Hardy 2002) have come into prominence as providing a rigorous foundation for subjectivist probability. Roughly speaking, this mathematical work shows that if the observer assessing subjective probabilities is to be logically consistent, then their plausibility estimates must obey the standard rules of probability.

From a philosophy-of-AI point of view, neither the frequentist nor the subjectivist interpretations, as commonly presented, is fully satisfactory. However, for reasons to be briefly explained here, we find the subjectivist interpretation more acceptable, and will consider indefinite probabilities within a subjectivist context, utilizing relative frequency calculations for pragmatic purposes but giving them an explicitly subjectivist rather than frequentist interpretation.

The frequentist interpretation is conceptually problematic in that it assigns probabilities only in terms of limits of sequences, not in terms of finite amounts of data. Furthermore, it has well-known difficulties with the assignment of probabilities to unique events that are not readily thought of as elements of ensembles. For instance, what was the probability, in 1999, of the statement S holding that "A great depression will be brought about by the Y2K problem"? Yes, this probability can be cast in terms of relative frequencies in various ways. For instance, one can define it as a relative frequency across a set of hypothetical "possible worlds": across all possible worlds similar to our own, in how many of them did the Y2K problem bring about a great depression? But it's not particularly natural to assume that this is what an intelligence must do in order to assign a probability to S. It would be absurd to claim that, in order to assign a probability to S, an intelligence must explicitly reason in terms of an ensemble of possible worlds. Rather, the claim must be that whatever reasoning a mind does to evaluate the probability of S may be implicitly interpreted in terms of possible worlds. This is not completely senseless, but is a bit of an irritating conceptual stretch.

The subjectivist approach, on the other hand, is normally conceptually founded either on rational betting behaviors or on Cox's theorem and its generalizations, both of which are somewhat idealistic.

No intelligent agent operating within a plausible amount of resources can embody fully self-consistent betting behavior in complex situations. The irrationality of human betting behavior is well known; to an extent this is due to emotional reasons, but there are also practical limitations on the complexity of the situation in which any finite mind can figure out the correct betting strategy.

And similarly, it is too much to expect any severely resource-constrained intelligence to be fully self-consistent in the sense that the assumptions of Cox's theorem require. In order to use Cox's theorem to justify the use of probability theory by practical intelligences, it seems to us, one would need to take another step beyond Cox, and argue that if an AI system is going to have a "mostly sensible" measure of plausibility (i.e., if its deviation from Cox's axioms are not too great), then its intrinsic plausibility measure must be *similar to* probability. We consider this to be a viable line of argument, but will pursue this point in another paper – to enlarge on such matters here would take us too far afield.

Walley's approach to representing uncertainty is based explicitly on a Bayesian, subjectivist interpretation; though whether his mathematics has an alternate frequentist interpretation is something he has not explored, to our knowledge. Similarly, our approach here is to take a subjectivist perspective on the foundational semantics of indefinite probabilities (although we don't consider this critical to our approach; quite likely it could be given a frequentist interpretation as well). Within our basic subjectivist interpretation, however, we will frequently utilize relative frequency calculations when convenient for pragmatic reasoning. This is conceptually consistent because within the subjectivist perspective there is still a role for relative frequency calculations, so long as they are properly interpreted.

Specifically, when handling a conditional probability $P(A|B)$, it may be the case that there is a decomposition $B=B_1+...+B_n$ so that the B_i are mutually exclusive and equiprobable, and each of $P(A|B_i)$ is either 0 or 1. In this case the laws of probability tell us $P(A|B) = P(A|B_1)\ P(B_1|\ B) + ... + P(A|B_n)\ P(B_n|B) = (P(A|B_1) + ... + P(A|B_n))/n$, which is exactly a relative frequency. So, in the case of statements that are decomposable in this sense, the Bayesian interpretation implies a relative frequency based interpretation (but not a "frequentist" interpretation in the classical sense). For decomposable statements, plausibility values may be regarded as the means of probability distributions, where the distributions may be derived via subsampling (sampling subsets C of $\{B_1,...,B_n\}$, calculating $P(A|C)$ for each subset, and taking the distribution of these values; as in the statistical technique known as bootstrapping). In the case of the "Y2K" statement and other similar statements regarding unique instances, one option is to think about decomposability across possible worlds, which is conceptually controversial.

4.4 Indefinite Probability

We concur with the subjectivist maxim that a probability can usefully be interpreted as an estimate of the plausibility of a statement, made by some observer. However, we suggest introducing into this notion a more careful consideration of the role of evidence in the assessment of plausibility. We introduce a distinction that we feel is critical, between

- the ordinary (or "definite") plausibility of a statement, interpreted as the degree to which the evidence already (directly or indirectly) collected by a particular observer supports the statement.
- the "indefinite plausibility" of a statement, interpreted as the degree to which the observer believes that the overall body of evidence potentially available to him supports the statement.

The indefinite plausibility is related to the ordinary plausibility, but also takes into account the potentially limited nature of the store of evidence collected by the observer at a given point in time. While the ordinary plausibility is effectively represented as a single number, the indefinite plausibility is more usefully represented in a more complex form. We suggest to represent an indefinite plausibility as a quadruple $<[L, U], b, k>$, which when attached to a statement *S* has the semantics "I assign an ordinary plausibility of *b* to the statement that 'Once *k* more items of evidence are collected, the ordinary plausibility of the statement *S* will lie in the interval [*L*, *U*]'." Note that indefinite plausibility is thus defined as "second-order plausibility" – a plausibility of a plausibility.

As we shall see in later sections of the paper, for most computational purposes it seems acceptable to leave the parameter k in the background, assuming it is the same for both the premises and the conclusion of an inference. So in the following we will speak mainly of indefinite probabilities as $<[L, U], b>$ triples, for sake of simplicity. The possibility does exist, however, that in future work inference algorithms will be designed that utilize k explicitly.

Now, suppose we buy the Bayesian argument that ordinary plausibility is best represented in terms of probability. Then it follows that indefinite plausibility is best represented in terms of second-order probability; i.e., as "I assign probability *b* to the statement that 'Once *k* more items of evidence have been collected, the probability of the truth of *S* based on this evidence will lie in the interval [*L*, *U*]'."

4.4.1 An Interpretation in Terms of Betting Behavior

To justify the above definition of indefinite probability more formally, one approach is to revert to betting arguments of the type made by de Finetti in his work on the foundations of probability. As will be expounded below, for computational purposes we have taken a pragmatic frequentist approach based on underlying distributional assumptions. However, for purposes of conceptual clarity, a more subjectivist de Finetti style justification is nevertheless of interest. So, in this subsection we will describe a "betting scenario" that leads naturally to a definition of indefinite probabilities.

Suppose we have a category C of discrete events; e.g., a set of tosses of a certain coin, which has heads on one side and tails on the other. Next, suppose we have a predicate *S*, which is either True or False (Boolean values) for each event within the above event-category C. For example, if C is a set of tosses of a certain coin, then *S* could be the event "Heads." *S* is a function from events into Boolean values.

If we have an agent A, and the agent A has observed the evaluation of *S* on n different events, then we will say that n is the amount of evidence that A has observed regarding *S*; or we will say that A has made n observations regarding *S*.

Now consider a situation with three agents: the House, the Gambler, and the Meta-gambler. As the name indicates, House is going to run a gambling operation, involving generating repeated events in category C, and proposing bets regarding the outcome of future events in C.

More interestingly, House is also going to propose bets to Meta-gambler regarding the behavior of Gambler.

Specifically, suppose House behaves as follows.

After Gambler makes *n* observations regarding *S*, House offers Gambler the opportunity to make what we'll call a "de Finetti" type bet regarding the outcome of the next observation of S. That is, House offers Gambler the opportunity:

> You must set the price of a promise to pay $1 if the next observation of S comes out True, and $0 if it does not. You must commit that I will be able to choose to either buy such a promise from you at the price you have set, or to require you to buy such a promise from me. In other words: you set the odds, but I decide which side of the bet will be yours.

Assuming Gambler does not want to lose money, the price Gambler sets in such a bet is the "operational subjective probability" that Gambler assigns that the next observation of *S* will come out True.

As an aside, House might also offer Gambler the opportunity to bet on sequences of observations; e.g., it might offer similar "de Finetti" price-setting opportunities regarding predicates like "The next 5 observations of *S* made will be in the ordered pattern (True, True, True, False, True)." In this case, things become interesting if we suppose Gambler thinks that: For each sequence *Z* of {True, False} values emerging from repeated observation of *S*, any permutation of *Z* has the same (operational subjective) probability as *Z*. Then, Gambler thinks that the series of observations of *S* is "exchangeable," which means intuitively that *S*'s subjective probability estimates are really estimates of the "underlying probability of *S* being true on a random occasion." Various mathematical conclusions follow from the assumption that Gambler does not want to lose money, combined with the assumption that Gambler believes in exchangeability.

Next, let's bring Meta-gambler into the picture. Suppose that House, Gambler and Meta-gambler have all together been watching *n* observations of *S*. Now, House is going to offer Meta-gambler a special opportunity. Namely, he is going to bring Meta-gambler into the back room for a period of time. During this period of time, House and Gambler will be partaking in a gambling process involving the predicate *S*.

Specifically, while Meta-gambler is in the back room, House is going to show Gambler *k* new observations of *S*. Then, after the *k*'th observation, House is going to come drag Meta-gambler out of the back room, away from the pleasures of the flesh and back to the place where gambling on *S* occurs.

House then offers Gambler the opportunity to set the price of yet another de Finetti style bet on yet another observation of *S*. Before Gambler gets to set his price, though, Meta-gambler is going to be given the opportunity of placing a bet regarding what price Gambler is going to set. Specifically, House is going to allow Meta-gambler to set the price of a de Finetti style bet on a proposition of Meta-gambler's choice, of the form:

Q = "Gambler is going to bet an amount p that lies in the interval [*L*, *U*]"

For instance Meta-gambler might propose

> Let Q be the proposition that Gambler is going to bet an amount lying in [.4, .6] on this next observation of *S*. I'll set at 30 cents the price of a promise defined as follows: To pay $1 if Q comes out True, and $0 if it does not. I will commit that you will be able to choose either to buy such a promise from me at this price, or to require me to buy such a promise from you.

I.e., Meta-Gambler sets the price corresponding to Q, but House gets to determine which side of the bet to take. Let us denote the price set by Meta-gambler as b; and let us assume that Meta-gambler does not want to lose money. Then, b is Meta-gambler's subjective probability assigned to the statement that:

> "Gambler's subjective probability for the next observation of *S* being True lies in [L, U]."

But, recall from earlier that the indefinite probability <[*L*, *U*], *b*, *k*> attached to *S* means that:

> "The estimated odds are b that after *k* more observations of *S*, the estimated probability of *S* will lie in [*L*, *U*]."

or in other words

> "[*L*, *U*] is a *b*-level credible interval for the estimated probability of *S* after *k* more observations."

In the context of an AI system reasoning using indefinite probabilities, there is no explicit separation between the Gambler and the Meta-gambler; the same AI system makes both levels of estimate. But this is of course not problematic, so long as the two components (first-order probability estimation and b-estimation) are carried out separately.

One might argue that this formalization in terms of betting behavior doesn't really add anything practical to the indefinite probabilities framework as already formulated. At minimum, however, it does make the relationship between indefinite probabilities and the classical subjective interpretation of probabilities quite clear.

4.4.2 A Pragmatic Frequentist Interpretation

Next, it is not hard to see how the above-presented interpretation of an indefinite plausibility can be provided with an alternate justification in relative frequency terms, in the case where one has a statement *S* that is decomposable in the sense described above. Suppose that, based on a certain finite amount of evidence about the frequency of a statement *S*, one wants to guess what one's frequency estimate will be once one has seen a lot more evidence. This guessing process will result in a probability distribution across frequency estimates – which may itself be interpreted as a frequency via a "possible worlds" interpretation. One may think about "the frequency, averaged across all possible worlds, that we live in a world in which the observed frequency of *S* after *k* more observations will lie in interval I." So, then, one may interpret <[*L*, *U*], *b*, *N*> as meaning "*b* is the frequency of possible worlds in which the observed frequency of *S*, after I've gathered *k* more pieces of evidence, will lie in the interval [*L*, *U*]."

This interpretation is not as conceptually compelling as the betting-based interpretation given above – because bets are real things, whereas these fictitious possible worlds are a bit slipperier. However, we make use of this frequency-based interpretation of indefinite probabilities in the practical computational implementation of indefinite probability presented in the following sections – without, of course, sacrificing the general Bayesian interpretation of the indefinite probability approach. In the end, we consider the various interpretations of probability to be in the main complementary rather than contradictory, providing different perspectives on the same very useful mathematics.

Moving on, then: To adopt a pragmatic frequency-based interpretation of the second-order plausibility in the definition of indefinite plausibility, we interpret "I assign probability *b* to the statement that 'Once *k* more items of evidence are collected, the probability of the truth of *S* based on this evidence will lie in the interval [*L*, *U*]'" to mean "*b* is the frequency, across all possible worlds in which I have gathered *k* more items of evidence about *S*, of worlds in which the statement 'the estimated probability of *S* lies in the interval [*L*, *U*]' is true." This frequency-based interpretation allows us to talk about a probability distribution consisting of probabilities assigned to values of 'the estimated probability of S,' evaluated across various possible worlds. This probability distribution is what, in the later sections of the paper, we call the "second-order distribution." For calculational purposes, we assume a particular distributional form for this second-order distribution.

Next, for the purpose of computational implementation, we make the heuristic assumption that the statement *S* under consideration is decomposable, so that in each possible world, "the estimated probability of *S*" may be interpreted as the mean of a probability distribution. For calculational purposes, in our current implementation we assume a particular distributional form for these probability distributions, which we refer to as "the first-order distributions."

The adoption of a frequency-based interpretation for the second-order plausibility seems hard to avoid if one wants to do practical calculations using the indefinite probabilities approach. On the other hand, the adoption of a frequency-based interpretation for the first-order plausibilities is an avoidable convenience, which is appropriate only in some situations. We will discuss below how the process of reasoning using indefinite probabilities can be simplified, at the cost of decreased robustness, in cases where decomposability of the first order probabilities is not a plausible assumption.

So, to summarize, in order to make the indefinite probabilities approach computationally tractable, we begin by restricting attention to some particular family D of probability distributions. Then, we interpret an interval probability attached to a statement as an assertion that: "There is probability *b* that the subjective probability of the statement, after I have made *k* more observations, will appear to be drawn from a distribution with a mean in this interval."

Then, finally, given this semantics and a logical inference rule, one can ask questions such as: "If each of the premises of my inference corresponds to some interval, so that there is probability *b* that after *k* more observations the distribution governing the premise will appear to have a mean in that interval, then what is an interval so that *b* of the family of distributions of the conclusion have means lying in that interval?" We may then give this final interval the interpretation that, after *k* more observations, there is a probability *b* that the conclusion of the inference will appear to lie in this final interval. (Note that, as mentioned above, the parameter *k* essentially "cancels out" during inference, so that one doesn't need to explicitly account for it during most inference operations, so long as one is willing to assume it is the same in the premises and the conclusion.)

In essence, this strategy merges the idea of imprecise probabilities with the Bayesian concept of credible intervals; thus the name "indefinite probabilities" ("definite" having the meaning of "precise," but also the meaning of "contained within specific boundaries" – Walley's probabilities are contained within specific boundaries, whereas ours are not).

As hinted above, however, the above descriptions mask the complexity of the actual truth-value objects. In the indefinite probabilities approach, in practice, each IndefiniteTruthValue object is also endowed with three additional parameters:

- An indicator of whether [L, U] should be considered as a symmetric or asymmetric credible interval.
- A family of "second-order" distributions, used to govern the second-order plausibilities described above.
- A family of "first-order" distributions, used to govern the first-order plausibilities described above.

Combined with these additional parameters, each truth-value object essentially provides a compact representation of a single second-order probability distribution with a particular, complex structure.

4.5 Truth-Value Conversions

In our current implementation, we usually use <[L, U], *b*, *k*> IndefiniteTruthValues for inference. For other purposes however, it is necessary to convert these truth values into SimpleTruthValues, in either (s,w) or (s,n) form. We now derive a conversion formula for translating indefinite truth values into simple truth values. In order to carry out the derivation additional assumptions must be made, which is why the formula derived here must be considered "heuristic" from the point of view of applications. When the underlying distributional assumptions apply, the formula is exact, but these assumptions may not always be realistic.

4.5.1 Calculation of Approximate Conversion Formulas

To derive conversion formulas we assume that the distributions underlying the means within the [L, U] intervals of the indefinite truth values are beta distributions. Due to the conjugacy of the beta and binomial distributions, this means we can model these means as corresponding to Bernoulli trials.

In order to derive approximate formulas, we first consider the problem "backwards": Given b, n, and k, how can we derive [L, U] from an assumption of an underlying Bernoulli process with unknown probability p? We then reverse the process to obtain an approach for deriving n given b, k and [L, U].

Theorem: Suppose there were x successes in the first n trials of a binomial process with an unknown fixed probability p for success. Suppose further that the prior probability density $f(p=a)$ is uniform. Then the probability that there will be $(x+X)$ successes in the first $(n+k)$ trials is given by

$$\frac{(n+1)\binom{k}{X}\binom{n}{x}}{(k+n+1)\binom{k+n}{X+x}},$$

where $\binom{}{}$ indicates the binomial coefficient.

Proof: The probability $P\left(y \text{ successes in } n+k \text{ trials} \middle| x \text{ successes in } n \text{ trials}\right)$ is the same as $P\left(y-x \text{ successes in } k \text{ trials}\right)$. Letting $X = y\text{-}x$ and assuming probability p for success on each trial, then this probability would be given by $\binom{k}{X}p^X(1-p)^{k-X}$.

The probability densities for p can be found from Bayes' theorem, which states

$$f\left(p = a \middle| s = \frac{x}{n}\right) = \frac{f\left(s = \frac{x}{n} \middle| p = a\right) f(p = a)}{f\left(s = \frac{x}{n}\right)}. \tag{1}$$

Here $f()$ denotes the appropriate probability density. Now $f(p=a)=1$ and since n is fixed, $f\left(s = \frac{x}{n}\right) = f(x \text{ successes in } n \text{ trials})$, so

$$f\left(s = \frac{x}{n}\right) = \int_0^1 \binom{n}{x} p^x (1-p)^{n-x} dp = \frac{1}{n+1}. \tag{2}$$

Hence,

$$\begin{aligned} &P(X \text{ successes in } k \text{ trials}) \\ &= (n+1)\int_0^1 \binom{k}{X}\binom{n}{x} p^{X+x} (1-p)^{k+n-(X+x)} dp \\ &= \frac{(n+1)\binom{k}{X}\binom{n}{x}}{(k+n+1)\binom{k+n}{X+x}}. \end{aligned} \tag{3}$$

The theorem gives a distribution based on n, k and x; and then, applying b, we can find a symmetric credible interval [L, U] about $s=x/n$ based on this distribution.

Due to small deviations arising from integer approximations, given L, U, b and k, the reverse process is somewhat trickier. We now outline two approximate inverse procedures. We first exhibit a heuristic algorithmic approach. From the results of this heuristic approach we then develop an approximate inverse function.

4.5.1.1 Heuristic Algorithmic Approach

1. Let $L_1 = \frac{\lfloor (n+k)L + 0.5 \rfloor}{\text{n}+\text{k}}$, and $U_1 = \frac{\lfloor (n+k)U + 0.5 \rfloor}{\text{n}+\text{k}}$.

2. Calculate $$b_1 = \frac{n+1}{k+n+1}\sum_{m=l}^{m=l+r}\frac{\binom{k}{m}\binom{n}{x}}{\binom{k+n}{m+x}},$$ where

$$x = \frac{\lfloor n(L+U)+0.5 \rfloor}{2}, \qquad l = \lfloor (n+k)L + 0.5 \rfloor, \qquad \text{and}$$
$$r = (n+k)(U_1 - L_1).$$

3. Form the function $v(n)=d[\langle[L1, U1], b1\rangle, \langle[L, U], b\rangle]$, where $d[a,b]$ is the standard Euclidean distance from a to b.
4. Find the value of n that minimizes the function $v(n)$.

Aside from small deviations arising from integer approximations, n depends inversely on the width U-L of the interval [L, U]. To find the n-value in step 4 we initially perform a search by setting $n=2^j$ for a sequence of j values, until we obtain a b value that indicates we have surpassed the correct n value. We then perform a binary search between this maximum value N and $N/2$. We thus guarantee that the actual algorithm is of order O(log n).

4.5.1.2 Approximate Inverse Function Approach

As an alternate and faster approach to finding n, we develop a function of L, U, k, and b that provides a reasonable approximation for n. We begin by plotting the cumulative probabilities given by equation (3) for various values of n, by following the first two steps of the heuristic approach above. Aside from small deviations caused by the discrete nature of the cumulative distribution functions, each graph can be approximately modeled by a function of the form $\text{prob} = 1 - b = 1 - Ae^{-Bn}$. Inverting, we obtain $n = -\frac{\ln\left[\frac{b}{A}\right]}{B}$.

For simplicity, we model the dependence of the coefficients A and B upon the values of L, U, and k, linearly. From the data we gathered this appears to be a reasonable assumption, though we have not yet derived an analysis of the error of these approximations. We will use the notation $A = A_1U + A_2$, $B = B_1U + B_2$, $A_i = A_{i1}L + A_{i2}$, $B_i = B_{i1}L + B_{i2}$, $A_{ij} = A_{ij1}k + A_{ij2}$, and $B_{ij} = B_{ij1}k + B_{ij2}$, where $1 \le i,j,k \le 2$. Putting everything together we end up with

$$A = (A_{11}L + A_{12})U + (A_{21}L + A_{22})$$
$$= \left[(A_{111}k + A_{112})L + A_{121}k + A_{122}\right]U + \left[(A_{211}k + A_{212})L + A_{221}k + A_{222}\right]$$

and

$$B = (B_{11}L + B_{12})U + (B_{21}L + B_{22})$$
$$= [(B_{111}k + B_{112})L + B_{121}k + B_{122}]U + [(B_{211}k + B_{212})L + B_{221}k + B_{222}]$$

Finding the values of the coefficients A_{ij} and B yields the following values:

A_{111} = -0.00875486	A_{112} = -2.35064019	A_{121} = 0.002463011
A_{122} = -0.220372781	A_{211} = 0.010727656	A_{212} = 2.803020516
A_{221} = -0.003647227	A_{222} = 0.437068392	
B_{111} = 0.003032946	B_{112} = -0.399778839	B_{121} = -0.004302594
B_{122} = 0.930153781	B_{211} = 0.002803518	B_{212} = -0.593689012
B_{221} = -0.000265616	B_{222} = -0.071902027	

Observing that the dependence upon k is relatively negligible compared to the dependence upon L, U, and b, we can alternatively eliminate the k-dependence and use instead fixed values for A_{11}=-2.922447948, A_{12}=-0.072074201, A_{21}=3.422902859, A_{22}=0.252879708, B_{11}=-0.261903438, B_{12}=0.716893418, B_{21}=-0.39831749, and B_{22}=-0.107482534.

4.5.2 Further Development of Indefinite Probabilities

In this chapter we have presented the basic idea of indefinite truth values. The purpose of the indefinite truth value idea, of course, lies in its utilization in inference, which is left for later chapters. But our hope is that in this chapter we have conveyed the essential semantics of indefinite probabilities, which is utilized, along the mathematics, in the integration of indefinite probabilities with inference rules. Some of the inferential applications of indefinite probabilities we encounter in later chapters will be fairly straightforward, such as their propagation through deductive and Bayesian inference. Others will be subtler, such as their application in the context of intensional or quantifier inference. In all cases, however, we have found the indefinite probability notion useful as a summary measure of truth value in a pragmatic inference context. In some cases, of course, a summary approximation won't do and one actually needs to retain one or more full probability distributions rather than just a few numbers giving a rough indication. But in reality one can't always use a full representation of this nature due to restrictions on data, memory, and processing power; and thus we have placed indefinite probabilities in a central role within PLN. As compared with simpler summary truth values such as single probability numbers or (probability, weight of evidence) pairs, they seem to provide a better compromise between compactness and accuracy.

Chapter 5: First-Order Extensional Inference – Rules and Strength Formulas

Abstract In this chapter we launch into the "meat" of PLN: the specific PLN inference rules, and the corresponding truth-value formulas used to determine the strength of the conclusion of an inference rule from the strengths of the premises. Inference rules and corresponding truth-value strength formulas comprise a large topic; in this chapter we deal with a particular sub-case of the problem: first-order extensional inference.

5.1 Introduction

Recall that first-order inference refers to inference on relationships between terms (rather than on predicates, or relationships between relationships), and that extensional inference refers to inference that treats terms as denoting sets with members (as opposed to intensional inference, which treats terms as denoting entities with properties). These first-order extensional rules and truth-value formulas turn out to be the core PLN rules truth-value formulas, in the sense that most of the rules and formulas for handling higher-order and/or intensional inference and/or weight of evidence are derived as re-interpretations of the first-order extensional rules and associated strength formulas. Higher-order inference, intensional inference, and weight of evidence formulas are handled separately in later chapters.

5.2 Independence-Assumption-Based PLN Deduction

In this section we present one version of the PLN strength formula for first-order extensional deduction (abbreviated to "deduction" in the remainder of the chapter). First we give some conceptual discussion related to this inference formula; then we provide the algebraic formula. The formula itself is quite simple; however it is important to fully understand the concepts underlying it, because it embodies simplifying assumptions that introduce errors in some cases. We will also be reviewing an alternate strength formula for first-order extensional deduction, in a later section of this chapter, which mitigates these errors in certain circumstances.

Conceptually, the situation handled by the first-order extensional deduction formula is depicted in the following Venn diagram:

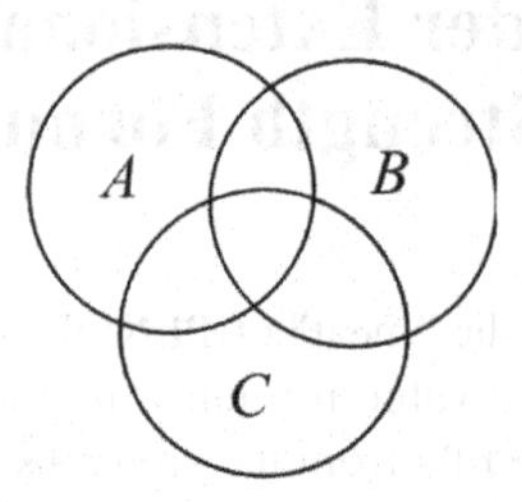

First-order extensional deduction, as well as the related inference forms we call induction and abduction in PLN, may be cast in the form: Given information about the size of some regions in the Venn diagram, make guesses about the size of other regions.

5.2.1 Conceptual Underpinnings of PLN Deduction

In this subsection, as a preliminary to presenting the PLN inference formulas in detail, we will discuss the concepts underlying PLN in a more abstract way, using simple inference examples to demonstrate what it is we really mean by "probabilistic deduction." This conceptual view is not needed for the actual calculations involved in PLN, but it's essential to understanding the semantics underlying these calculations, and is also used in the proof of the PLN deduction formula.

Let's consider some simple examples regarding Subset relationships. Supposing, in the above diagram, we know

```
Subset A B <.5>
Subset B C <.5>
```

What conclusions can we draw from these two relationships? What we want is to derive a relation of the form

```
Subset A C <tv>
```

from the two given premises. When we do this however, we are necessarily doing probabilistic, estimative inference, not direct truth-value evaluation.

To see why, suppose *B* is a set with two elements; e.g.,

$$B = \{x_1, x_2\}$$

so that

Member x_1 *B* <1>
Member x_2 *B* <1>

Let's consider two of the many possible cases that might underlie the above Subset relationships:

Case 1 ($A \cap B$ and $B \cap C$ are identical)

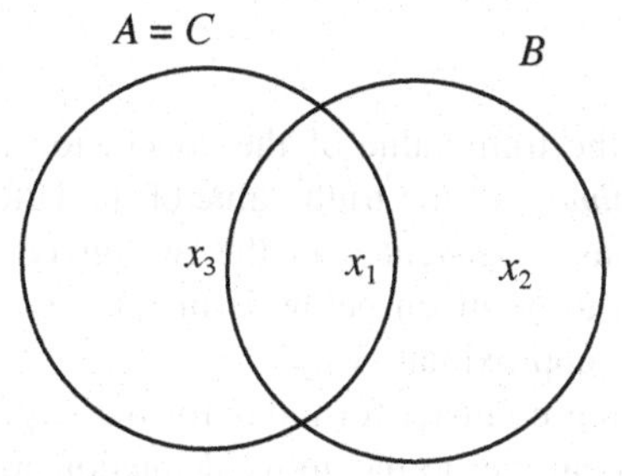

$A = C = \{x_1, x_3\}$
$B = \{x_1, x_2\}$

In this case, direct truth-value evaluation yields

```
Subset A C <1>
```

Case 2 ($A \cap B$ and $B \cap C$ are disjoint)

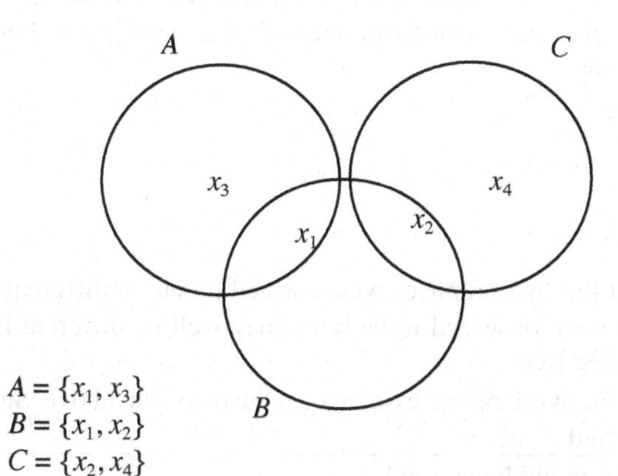

$A = \{x_1, x_3\}$
$B = \{x_1, x_2\}$
$C = \{x_2, x_4\}$

In this case, direct truth-value evaluation yields

```
Subset A C <0>
```

The problem is that, just given the two premises

```
Subset A B <.5>
Subset B C <.5>
```

we don't know which of the two above cases holds – or if in fact it's a completely different situation underlying the Subset relationships. Different possible situations, underlying the same pair of Subset relationships, may result in completely different truth-values <tv> for the conclusion

```
Subset A C <tv>
```

So no exact computation of the truth-value of the conclusion is possible. Instead, all that's possible is an *estimate* of the truth-value of the latter, obtained by averaging over possible situations consistent with the two given premises. The PLN deduction strength formula to be given below is merely an *efficient way of carrying out this averaging,* in an approximate way.

We will derive theorems giving compact formulas for the expected (average) truth-values to be obtained as the answer to the above deduction problem. We approach this averaging process in two ways. First, we derive inference formulas for an "independence-based PLN," in which we average over all possible sets satisfying given premises. Later, we introduce a more general and more realistic "concept-geometry" approach. In the concept-geometry approach, we restrict attention to only those sets having particular shapes. The idea here is that concepts tend to be better approximated by these shapes than by random sets.

One important point regarding PLN inference is that the *evidence sets used* to compute the truth-values of two premises need not be overlapping. We can freely combine two relationships that were derived based on different sets of observations. This is important because many real-world cases fit this description. For instance, when a person reasons

```
Inheritance gulls birds
Inheritance birds fly
|-
Inheritance gulls fly
```

it may well be the case that the two premises were formed based on different sets of evidence. The gulls that were observed to be birds may well be different from the birds that were observed to fly.

A simple example of non-overlapping evidence, related to our above Subset example with the sets *A*, *B*, and *C*, is:

Case 3 (non-overlapping evidence sets)

$A = \{x_1, x_4, x_7, x_8\}$
$B = \{x_1, x_2, x_3, x_4, x_5, x_6\}$
$C = \{x_2, x_3, x_4\}$

In this case, direct truth-value evaluation yields

```
Subset A C <.25>
```

Now, suppose that truth-value estimates for the two relations

```
Subset A B <.5>
Subset B C <.5>
```

were obtained as follows:

```
Subset A B <.5>
```

was derived by observing only $\{x_1, x_7\}$, whereas

```
Subset B C <.5>
```

was derived by observing only $\{x_2, x_6\}$

In Case 3, the two premise Subsets were derived based on completely different sets of evidence. But the inference rules don't care; they can make their estimates anyway. We will present this "not caring" in a formal way a little later when we discuss the weight of evidence rules for first-order inference.

Now let's step through the basic logic by which PLN deals with these examples of inference involving Subsets, with all their concomitant subcases. Basically, as noted above, what PLN does is to *estimate* the outcome of an average over all possible cases. The PLN formulas carry out this estimation in a generally plausible way, based on the information available.

More formally, what does this mean? First we'll present a slight simplification, then un-simplify it in two stages. In addition to these simplifications, we will also assume a strength-only representation of truth-values, deferring consideration of more complex truth-value types until later.

In the simplified version what PLN says in this example is, basically: Let's assume a universe U that is a finite size Usize. Let V denote the set of all triples of sets $\{A, B, C\}$ in this universe, such that

```
Subset A B <.5>
Subset B C <.5>
```

holds. For each triple in V, we can compute the value s so that

```
Subset A C <s>
```

The average of all these strength-estimates is the estimated truth-value strength of the Subset. Note that this average includes cases where $A \cap B$ and $B \cap C$ have no overlap at all, cases where $A \cap B$ and $B \cap C$ are identical, cases where A and C are identical – *all possible cases* given the assumed finite universe U and the constraints posed by the premise Subsets. If the two premise Subsets are drawn from different evidence sets, this doesn't matter – the degree of evidence-set overlap is just one among many unknown properties of the evidence sets underlying the premise truth-values.

Now, how is this picture a simplification? The first way is that we haven't introduced all the potentially available information. We may have knowledge about the truth-values of *A*, *B*, and *C*, as well as about the truth-values of the Subset relationships. In fact, this is the usual case. Suppose that s_A, s_B, and s_C are the strengths of the truth-values of the Atoms *A*, *B*, and *C*. In that case, we can redefine the set *V* specified above; we can define it as the set of all triples of sets $\{A, B, C\}$ so that

```
|A| = sA Usize
|B| = sB Usize
|C| = sC Usize
Subset A B <.5>
Subset B C <.5>
```

hold. Here Usize is the "universe size," the size of the total space of entities under consideration. We can then compute the truth-value of

```
Subset A C <tv>
```

as the average of the estimates obtained for all triples in *V*.

Next, the *other* way in which the above picture is a simplification is that it assumes the strength values of the premises (and the strength values of the Atoms *A*, *B*, and *C*) are exactly known. In fact, these will usually be estimated values; and if we're using indefinite truth-values, distributional truth-values, or confidence-bearing truth-values, knowledge about the "estimation error" may be available. In these cases, we are not simply forming a set *V* as above. Instead, we are looking at a probability distribution *V* over all triples $\{A, B, C\}$ where *A*, *B*, and *C* are subsets of *U*, and the quantities

```
|A| (determined from sA = A.TruthValue.strength)
|B| (determined from sB = B.TruthValue.strength)
|C| (determined from sC = C.TruthValue.strength)
(Subset A B).TruthValue.strength
(Subset B C).TruthValue.strength
```

are drawn from the probability distributions specified by the given truth-values. We will deal with this uncertainty below by doing a sensitivity analysis of the PLN inference formulas, estimating for each formula the error that may ensue from uncertain inputs, and discounting the weight of evidence associated with the conclusion based on this estimated error.

Finally, one technical point that comes up in PLN is that the quantitative results of truth-value estimates depend on the finite universe size Usize that is assumed. This parameter is also called the "context size." Basically, the smaller Usize is, the more speculative the inferences are. In the example given above, the minimum usable Usize is Usize = $|U| = |A| + |B| + |C|$. This parameter setting is good if one wants to do speculative inductive or abductive inference. If one wants to minimize error, at cost of also minimizing creativity, one should set Usize as large as possible. Using the PLN formulas, there is no additional computational cost in assuming a large Usize; the choice of a Usize value can be based purely on the desired inferential adventurousness. The semantics of the universal set will be discussed further in a later subsection.

5.2.2 The Independence-Assumption-Based Deduction Rule

Now we proceed to give a heuristic derivation of one of the two truth-value strength formulas commonly associated with the PLN deduction rule. In this inference rule, we want to compute the "strength" s_{AC} of the relationship

```
Subset A C <sAC>
```

which we interpret as

$$s_{AC} = P(C|A) = \frac{P(A \cap C)}{P(A)}$$

given the data:

$$s_{AB} = P(B|A)$$
$$s_{BC} = P(C|B)$$
$$s_A = P(A)$$
$$s_B = P(B)$$
$$s_C = P(C)$$

Essentially, that means we want to guess the size of the Venn diagram region $A \cap C$ given information about the sizes of the other regions A, B, C, $A \cap B$, $B \cap C$.

As illustrated in the above example, in the following discussion we will sometimes use the notation:

- s_{ij}'s are strengths of Subset relationships (e.g., Subset *i j* $<s_{ij}>$)
- s_i's are strengths of terms (i.e., *i* $<s_i>$)

This notation is handy for the presentation of algebraic relationships involving Atom strengths, a situation that often arises with PLN. We will also use N_{ij} and N_i to denote relationship and term "counts" respectively, and d_{ij} and d_j to denote relationship and term "weights of evidence."

Whenever the set of values $\{s_A, s_B, s_C, s_{AB}, s_{BC}\}$ is consistent (i.e., when it corresponds to some possible sets *A*, *B*, and *C*) then the PLN deduction strength formula becomes

$$s_{AC} = s_{AB}s_{BC} + \frac{(1-s_{AB})(s_C - s_B s_{BC})}{1-s_B} \quad \textbf{(2.1)}$$

Here we will give a relatively simple heuristic proof of this formula, and then (more tediously) a fully rigorous demonstration.

As we shall see a little later, formulas for inductive and abductive reasoning involve similar problems, where one is given some conditional and absolute probabilities and needs to derive other, indirectly related ones. For induction, one starts with s_{BA}, s_{BC}, s_A, s_B, s_C and wishes to derive s_{AC}. For abduction, one starts with s_{AB}, s_{CB}, s_A, s_B, s_C and wants to derive s_{AC}. The inference formulas involving similarity relations may be similarly formulated.

5.2.2.1 Heuristic Derivation of the Independence-Assumption-Based Deduction Rule

The heuristic derivation that we'll give here relies on a heuristic independence assumption. The rigorous derivation given afterward replaces the appeal to an independence assumption with an averaging over all possible worlds consistent with the constraints given in the premises. But of course, even the rigorous proof embodies some a priori assumptions. It assumes that the only constraints are the ones implicitly posed by the truth-values of the premise terms and relationships, and that every possible world consistent with these truth-values is equally likely. If there is knowledge of probabilistic dependency, this constitutes a bias on the space of possible worlds, which renders the assumption of unbiased independence invalid. Knowledge of dependency can be taken into account by modifying the inference formula, in a way that will be discussed below. The danger is where there is a dependency that is unknown; in this case the results of inference will not be accurate, an unavoidable problem.

Assume we're given $P(A)$, $P(B)$, $P(C)$, $P(B|A)$, $P(C|B)$, defined relative to some universal set U. We want to derive $P(C|A)$; i.e., the formula for $P(C|A) = s_{AC}$, which was cited above. We begin the derivation by observing that

$$P(C|A) = \frac{P(C \cap A)}{P(A)}.$$

Because $P(A)$ is assumed given, we may focus on finding $P(C \cap A)$ in terms of the other given quantities.

We know

$$P(C \cap A) = P(B)P((C \cap A)|B) + P(U - B)P((C \cap A)|U - B).$$

This follows because in general

$$\begin{aligned}
&P(B)P(X|B) + P(U - B)P(X|U - B) \\
&= \frac{P(B)P(X \cap B)}{P(B)} + \frac{P(U - B)P(X \cap (U - B))}{P(U - B)} \\
&= P(X \cap B) + P(X \cap (U - B)) \\
&= P(X \cap B) + [P(X) - P(X \cap B)] = P(X).
\end{aligned}$$

So we can say, for instance,

$$\begin{aligned}
P((C \cap A)|B) &= \frac{P(C \cap A \cap B)}{P(B)} \\
&= \frac{P((C \cap B) \cap (A \cap B))}{P(B)} \\
&= \frac{|(C \cap B) \cap (A \cap B)|}{|B|}.
\end{aligned}$$

Now, we can't go further than this without making an independence assumption. But if we assume C and A are independent (in both B and U-B), we can simplify these terms.

To introduce the independence assumption heuristically, we will introduce a "bag full of balls" problem. Consider a bag with N balls in it, b black ones and w white ones. We are going to pick n balls out of it, one after another, and we want to know the chance $p(k)$ that k of them are black. The solution to this problem is known to be the hypergeometric distribution. Specifically,

$$p(k) = \binom{b}{k}\binom{N-b}{n-k} \Big/ \binom{N}{n}.$$

The mean of this distribution is:

$$\text{mean} = \frac{bn}{N}.$$

How does this bag full of balls relate to our situation? We may say:

- Let our bag be the set B, so $N=|B|$.
- Let the black balls in the bag correspond to the elements of the set $A \cap B$, so that $b=|A \cap B|$
- The white balls then correspond to elements of $B\text{-}A$.
- The n balls we are picking are the elements of the set $C \cap B$, so $n=|C \cap B|$.

This probabilistic "bag full of balls" model embodies the assumption that A and C are totally independent and uncorrelated, so that once B and A are fixed, the chance of a particular subset of size $|C \cap B|$ lying in $A \cap B$ is the same as the chance of that element of C lying in a randomly chosen set of size $|A \cap B|$.

This yields the formula:

$$\frac{bn}{N} = \frac{|A \cap B||C \cap B|}{|B|}$$

which is an estimate of the size

$$|(C \cap B) \cap (A \cap B)|.$$

So if we assume that A and C are independent inside B, we can say

$$P((C \cap A)|B) = \frac{|(C \cap B) \cap (A \cap B)|}{|B|}$$

$$= \frac{|A \cap B||C \cap B|}{|B|^2} = P(A|B)P(C|B).$$

Similarly, for the second term, by simply replacing B with $U\text{-}B$ and then doing some algebra, we find

$$P((C \cap A)|(U-B)) = P(A|(U-B))P(C|(U-B))$$
$$= \frac{|A \cap (U-B)||C \cap (U-B)|}{|U-B|^2}$$
$$= \frac{[P(A)-P(A \cap B)][P(C)-P(A \cap B)]}{[1-P(B)]^2}.$$

So altogether, we find

$$P(C \cap A) = P(B)P((C \cap A)|B) + P(U-B)P((C \cap A)|(U-B))$$
$$= P(B)P(A|B)P(C|B) + \frac{[1-P(B)][P(A)-P(A \cap B)][P(C)-P(C \cap B)]}{[1-P(B)]^2}$$

and hence

$$P(C|A) = \frac{P(C \cap A)}{P(A)}$$
$$= \frac{P(A|B)P(C|B)P(B)}{P(A)} + \frac{[1-P(A \cap B)][P(C)-P(C \cap B)]}{1-P(B)}$$
$$= P(B|A)P(C|B) + \frac{[1-P(B|A)][P(C)-P(C \cap B)]}{1-P(B)}$$
$$= P(B|A)P(C|B) + \frac{[1-P(B|A)][P(C)-P(B)P(C|B)]}{1-P(B)}.$$

Note that in the above we have used Bayes' rule to convert

$$P(A|B) = \frac{P(B|A)P(A)}{P(B)}.$$

We now have the PLN deduction formula expressed in terms of conditional and absolute probabilities. In terms of our above-introduced notation for term and relationship strengths, we may translate this into:

$$s_{AC} = s_{AB} s_{BC} + \frac{(1 - s_{AB})(s_C - s_C s_{BC})}{1 - s_B}$$

which is the formula mentioned above.

5.2.2.2 PLN Deduction and Second-Order Probability

We now give a formal proof of the independence-assumption-based PLN deduction formula. While the proof involves a lot of technical details that aren't conceptually critical, there is one aspect of the proof that sheds some light on the more philosophical aspects of PLN theory: this is its use of second-order probabilities.

We first define what it means for a set of probabilities to be consistent with each other. Note that, given specific values for $s_A = P(A)$ and $s_B = P(B)$, not all values in the interval [0,1] for $s_{AB} = P(B|A)$ necessarily make sense. For example, if $P(A) = 0.6 = P(B)$, then the minimum value for $P(A \cap B) = 0.2$ so that the minimum value for $P(B|A)$ is $0.2/0.6 = 1/3$.

Definition: We say that the ordered triple (s_A, s_B, s_{AB}) of probability values $s_A = P(A)$, $s_B = P(B)$, and $s_{AB} = P(B|A)$ is consistent if the probabilities satisfy the following condition:

$$\max\left(0, \frac{s_A + s_B - 1}{s_A}\right) \le s_{AB} \le \min\left(1, \frac{s_B}{s_A}\right).$$

Definition: The ordered triple of subsets (A, B, C) for which the ordered triples (s_A, s_B, s_{AB}) and (s_B, s_C, s_{BC}) are both consistent, we shall call fully consistent subset-triples.

We will prove:

Theorem 1 (PLN Deduction Formula)

Let U denote a set with |U| elements. Let Sub(m) denote the set of subsets of U containing m elements. Let (A, B, C) be a fully consistent subset-triple. Further suppose that each of the values s_A, s_B, s_C, s_{AB}, and s_{BC} divides evenly into |U|. Next, define

$$f(x) = P\left[P(C|A) = x \middle| A \in Sub(|U| s_A), B \in Sub(|U| s_B), C \in Sub(|U| s_C), P(B|A) = s_{AB}, P(C|B) = s_{BC}\right]$$

Then, where E() denotes the expected value (mean), we have

$$E\left[f(x)\right] = s_{AC} = s_{AB}s_{BC} + \frac{(1-s_{AB})(s_C - s_B s_{BC})}{1-s_B}.$$

This theorem looks at the space of all finite universal sets U (all "sample spaces"), and with each U it looks at all possible ways of selecting subsets A, B, and C out of U. It assumes that this size of U is given, and that certain absolute and conditional probabilities regarding A, B, and C are given. Namely, it assumes that $P(A)$, $P(B)$, $P(C)$, $P(B|A)$ and $P(C|B)$ are given, but not $P(C|A)$. For each U, it then looks at the average over all A, B, and C satisfying the given probabilities, and asks: If we average across all the pertinent (A, B, C) triples, what will $P(C|A)$ come out to, on average? Clearly, $P(C|A)$ may come out differently for different sets A, B, and C satisfying the assumed probabilities. But some values are more likely than others, and we're looking for the mean of the distribution of $P(C|A)$ values over the space of acceptable (A, B, C) triples. This is a bit different from the usual elementary-probability theorems in that it's a second-order probability: we're not looking at the probability of an event, but rather the mean of a probability over a certain set of sample spaces (sample spaces satisfying the initially given probabilistic relationships).

In spite of the abstractness induced by the use of second-order probabilities, the proof is not particularly difficult. Essentially, after one sets up the average over pertinent sample spaces and does some algebra, one arrives at the same sort of hypergeometric distribution problem that was used in the heuristic derivation in the main text. The difference, however, is that in this proof there is no ad-hoc independence assumption; rather, the independence comes out of the averaging process automatically because on average, approximate probabilistic independence between terms is the rule, not the exception.

Proof of Theorem 1 (PLN Deduction Formula):

The way the theorem is stated, we start with a set U of $|U|$ elements, and we look at the set of all subset-triples $\{A, B, C\}$ fulfilling the given constraints. That is, we are looking at subset-triples (A,B,C) for which the Predicate *constr* defined by

$$constr(A,B,C) = A \in Sub(|U|s_A) \text{ AND } B \in Sub(|U|s_B) \text{ AND } C \in Sub(|U|s_C)$$
$$\text{AND } P(B|A) = s_{AB} \text{ AND } P(C|B) = s_{BC}$$

evaluates to True.

Over this set of subset-triples, we're computing the average of $P(C|A)$. That is, we're computing

$$E[f(x)] = \frac{1}{M}\sum_{(A,B,C)\,:\,constr(A,B,C)} P(C|A)$$

where M denotes the number of triples (A,B,C) so that $constr(A,B,C)$.

Following the lines of our heuristic derivation of the formula, we may split this into two sums as follows:

$$E[f(x)] = \frac{1}{M}\sum_{(A,B,C)\,:\,constr(A,B,C)} \frac{P(C\cap A)}{P(A)}$$

$$= \frac{1}{M}\left[\sum_{(A,B,C)\,:\,constr(A,B,C)} \frac{P((C\cap A)|B)P(B)}{P(A)} + \sum_{(A,B,C)\,:\,constr(A,B,C)} \frac{P((C\cap A)|(U-B))P(U-B)}{P(A)}\right].$$

After going this far, the heuristic derivation then used probabilistic independence to split up $P((C\cap A)|B)$ and $P(C\cap A)|(U-B)$ into two simpler terms apiece. Following that, the rest of the heuristic derivation was a series of straightforward algebraic substitutions. Our task here will be to more rigorously justify the use of the independence assumption. Here we will not make an independence assumption; rather, the independence will be implicit in the algebra of the summations that are "summations over all possible sets consistent with the given constraints." We will use formal methods analogous to the heuristic independence assumption, to reduce these sums into simple formulas consistent with the heuristic derivation.

We will discuss only the first sum here; the other one follows similarly by substituting $U-B$ for B. For the first sum we need to justify the following series of steps:

$$\frac{1}{M}\sum_{(A,B,C)\,:\,constr(A,B,C)} \frac{P((C\cap A)|B)P(B)}{P(A)}$$

$$= \frac{1}{M}\sum_{(A,B,C)\,:\,constr(A,B,C)} \frac{P(C\cap A\cap B)}{P(A)}$$

$$= \frac{1}{M}\sum_{(A,B,C)\,:\,constr(A,B,C)} \frac{P((C\cap B)\cap(A\cap B))}{P(A)}$$

$$= \frac{1}{M}\sum_{(A,B,C)\,:\,constr(A,B,C)} \frac{P(C\cap B)P(A\cap B)}{P(B)P(A)}$$

$$= \frac{1}{M}\sum_{(A,B,C)\,:\,constr(A,B,C)} \frac{P(C|B)P(A|B)P(B)}{P(A)}$$

The final step is the elegant one. It follows because, over the space of all triples (A,B,C) so that $constr(A,B,C)$ holds, the quantities $P(C|B)$ and $P(B|A)$ are constant by assumption. So they may be taken out of the summation, which has exactly M terms.

The difficult step to justify is the third one, where we transform $P((C \cap B) \cap (A \cap B))$ into $\frac{P(C \cap B)P(A \cap B)}{P(B)}$. This is where the algebra of the summations is used to give the effect of an independence assumption.

To justify this third transformation, it suffices to show that

$$\sum_{(A,B,C):\, constr(A,B,C)} \frac{P((C \cap B) \cap (A \cap B)) - \frac{P(C \cap B)P(A \cap B)}{P(B)}}{P(A)} = 0.$$

We will do this by rewriting the sum as

$$\sum_{B \in Sub(|U|\, sB)} \left[\sum_{(A,C):\, constr(A,B,C)} \left[\frac{P((C \cap B) \cap (A \cap B)) - \frac{P(C \cap B)P(A \cap B)}{P(B)}}{P(A)} \right] \right] = 0 \cdot$$

Note that the term $P(A)$ is constant for all (A,B,C) satisfying $constr(A,B,C)$, so it may be taken out of the summation and effectively removed from consideration, yielding

$$\sum_{B \in Sub(|U|\, sB)} \left[\sum_{(A,C):\, constr(A,B,C)} \left[P((C \cap B) \cap (A \cap B)) - \frac{P(C \cap B)P(A \cap B)}{P(B)} \right] \right] = 0 \cdot$$

We will show that this is true by showing that the inner summation itself is always zero; i.e., that for a fixed B,

$$\sum_{(A,C):\, constr(A,B,C)} \left[P((C \cap B) \cap (A \cap B)) - \frac{P(C \cap B)P(A \cap B)}{P(B)} \right] = 0 \qquad \textbf{(2.2)}$$

In order to demonstrate Equation 2.2, we will now recast the indices of summation in a different-looking but equivalent form, changing the constraint to one that makes more sense in the context of a fixed B.

Given a fixed B, let's say that a pair of sets $(A1, C1)$ is ***B-relevant*** iff it satisfies the relationships

$A1 = A \cap B$
$C1 = C \cap B$

for ***some*** triple (A,B,C) satisfying $constr(A,B,C)$.

We now observe that the pair $(A1,C1)$ is B-relevant if it satisfies the constraint predicate

$$constr1(A1,C1;B) = A1 \subseteq B \text{ AND } C1 \subseteq B \text{ AND } |A1| = |A| s_{AB} \text{ AND } |C1| = |B| s_{BC}$$

The constraint for $|C1|$ comes from the term in the former constraint *constr* stating

$$P(C|B) = s_{BC}$$

For, we may reason

$$P(C|B) = \frac{P(C \cap B)}{P(B)} = \frac{P(C1)}{P(B)} = s_{BC}$$

$$P(C1) = P(B) s_{BC}$$

$$|C1| = |B| s_{BC}$$

Similarly, to get the constraint for $|A1|$, we observe that

$$P(B|A) = \frac{P(A \cap B)}{P(A)} = \frac{P(A1)}{P(A)} = s_{AB}$$

so that

$$|A1| = |A| s_{AB}.$$

Given a fixed B, and a specific B-relevant pair $(A1,C1)$, let $EQ(A1, C1;B)$ denote the set of pairs (A,C) for which $constr(A,B,C)$ *and*

$$A1 = A \cap B$$

$$C1 = C \cap B$$

Now we will recast Equation 2.2 above in terms of A1 and C1. Equation 2.2 is equivalent to

$$\sum_{(A1,C1):\, constr1(A1,C1;B)} \sum_{(A,C) \in EQ(A1,C1;B)} \left[P(A1 \cap C1) - \frac{P(A1)P(C1)}{P(B)} \right] = 0.$$

Because the inner sum is the same for each pair $(A,C) \in \mathrm{EQ}(A1,C1;B)$, because $K \equiv |\mathrm{EQ}(A1,C1)|$ is the same for each B-relevant pair $(A1,C1)$, we can rewrite this as

$$\sum_{(A1,C1):constr1(A1,C1;B)} \left[K \left[P(A1 \cap C1) - \frac{P(A1)P(C1)}{P(B)} \right] \right] = 0$$

or just

$$\sum_{(A1,C1):constr1(A1,C1;B)} \left[P(A1 \cap C1) - \frac{P(A1)P(C1)}{P(B)} \right] = 0. \qquad \textbf{(2.3)}$$

We now have a somewhat simpler mathematics problem. We have a finite set B, with two subsets $A1$ and $C1$ of known sizes. Other than their sizes, nothing about $A1$ and $C1$ is known. We need to sum a certain quantity

$$Q \equiv P(A1 \cap C1) - \frac{P(A1)P(C1)}{P(B)}$$

over all possibilities for $A1$ and $C1$ with the given fixed sizes. We want to show this sum comes out to zero. This is equivalent to showing that the average of Q is zero, over all $A1$ and $C1$ with the given fixed sizes.

Now, the second term of Q is constant with respect to averaging over pairs $(A1,C1)$, because

$$\frac{P(A1)P(C1)}{P(B)} = \frac{|A1||C1|}{|B||U|}$$

which is independent of what the sets $A1$ and $C1$ are, assuming they have fixed sizes. So the average of the second term is simply $\frac{|A1||C1|}{|B||U|}$. We will rewrite Equation 2.3 as

$$L \equiv \frac{1}{M1} \sum_{A1 \subseteq B : |A1| = |A| s_{AB}} \left[\frac{1}{M2} \sum_{C1 : constr1(A1,C1;B)} P(A1 \cap C1) \right] = \frac{|A1||C1|}{|B||U|} \qquad \textbf{(2.4)}$$

where

- $M1$ is the number of $A1$'s that serve as part of a B-relevant pair $(A1,C1)$ with *any* $C1$; i.e., the number of terms in the outer sum;
- $M2$ is the number of $C1$'s that serve as part of a B-relevant pair $(A1,C1)$ for some specific $A1$; i.e., the number of terms in the inner sum.

Note that $M2$ is independent of the particular set $A1$ under consideration; and that $M = M1\ M2$.

To show (2.4), it suffices to show that for a fixed $A1$,

$$\frac{1}{M2}\sum_{C1:\,constr1(A1,C1;B)} P(A1 \cap C1) = \frac{|A1||C1|}{|B||U|} \tag{2.5}$$

To see why this suffices observe that, by the definition in (2.4), if (2.5) held, we'd have

$$L \equiv \frac{1}{M1}\sum_{A1 \subseteq B:|A1|=|A|s_{AB}} \left[\frac{|A1||C1|}{|B||U|}\right].$$

But the expression inside the sum is constant for all $A1$ being summed over (because they all have $|A1| = |A|s_{AB}$), and the number of terms in the sum is $M1$, so that on the assumption of (2.4) we obtain the result

$$L \equiv \frac{|A1||C1|}{|B||U|}.$$

which is what (2.3) states.

So, our task now is to show (2.5). Toward this end we will use an equivalent form of (2.4); namely

$$\frac{1}{M2}\sum_{C1:\,constr1(A1,C1;B)} |A1 \cap C1| = \frac{|A1||C1|}{|B|} \tag{2.6}$$

(the equivalence follows from $P(A1 \cap C1) = |A1 \cap C1|/|U|$). To show (2.6) we can use some standard probability theory, similar to the independence-assumption-based step in the heuristic derivation. We will model the left-hand side of (2.6) as a "bag full of balls" problem. Consider a bag with I balls in it, I black ones and w white ones. We are going to pick n balls out of it, one after another, and we want to know the chance $p(k)$ that k of them are black. The solution to this problem is known to be the hypergeometric distribution, as given above, with mean bn/N.

How does this bag full of balls relate to (2.6)? Simply:

- Let our bag be the set B, so $N=|B|$.
- Let the black balls in the bag correspond to the elements of the set $A1$, so that $b=|A1|$
- The white balls then correspond to B minus the elements in $A1$.
- The n balls we are picking are the elements of the set $C1$, so $n=|C1|$.

This yields the formula:

$$\frac{bn}{N} = \frac{|A1||C1|}{|B|}.$$

What the mean of the hypergeometric distribution gives us is the average of $|A1 \cap C1|$ over all I1 with the given size constraint, for a fixed $A1$ with the given size constraint.

But what Equation 5 states is precisely that this mean is equal to $\frac{|A1||C1|}{|B|}$. So, going back to the start of the proof, we have successfully shown that

$$\sum_{(A,B,C):constr(A,B,C)} \frac{P((C \cap A)|B)}{P(B)} = P(C|B)P(B|A).$$

It follows similarly that

$$\sum_{(A,B,C)constr(A,B,C)} \frac{P((C \cap A)(U-B))}{P(B)} = P(C|(U-B))P((U-B)|A).$$

The algebraic transformations made in the heuristic derivation then show that

$$E[f(x)] = \frac{1}{M}\sum_{\{A,B,C\}:\, constr(A,B,C)} P(C|A)$$

$$= P(C|B)P(B|A) + \frac{P(C|(U-B))P((U-B)|A)}{P(U-B)}$$

$$= P(B|A)P(C|B) + \frac{(1-P(B|A))(P(C)-P(B)P(C|B))}{1-P(B)}$$

which, after a change from P to s notation, is precisely the formula given in the theorem. Thus the proof is complete. **QED**

Next, to shed some light on the behavior of this formula, we now supply graphical plots for several different input values. These plots were produced in the Maple software package. Each plot will be preceded by the Maple code used to generate it. To make the Maple code clearer we will set Maple inputs in bold text; e.g.,

diff(dedAC(rB,rC,sAB,sBC),rC);

and Maple outputs through displayed text such as

$$\frac{(1 - sAB)}{1 - sB}$$

Recall that in order to apply the deduction rule, the triple (A,B,C) of subsets of a given universal set must be fully consistent. In Maple, this consistency condition takes the form

consistency:= (sA, sB, sC, sAB, sBC) -> (Heaviside(sAB-max(((sA+sB-1)/sA),0))-Heaviside(sAB-min(1,(sB/sA))))*(Heaviside(sBC-max(((sB+sC-1)/sB),0))-Heaviside(sBC-min(1,(sC/sB))));

where Heaviside(x) is the Heaviside unit step function defined by

$$\text{Heaviside}(x)=\begin{cases} 0, & \text{if } x < 0 \\ 1, & \text{if } x \geq 0 \end{cases}$$

The deduction strength formula then becomes

dedAC := (sA, sB, sC, sAB, sBC) -> sAB * sBC + (1- sAB)*(sC-sB*sBC)/(1-sB)* consistency(sA,sB,sC,sAB,sBC);

The result of this treatment of the consistency condition is that when the consistency condition indicates inconsistency, the result of the inference comes out as zero. Graphically, this means that the graphs look flat (0 on the z-axis) in certain regions – these are regions where the premises are inconsistent.

We now supply a sampling of graphs for the deduction strength formula for several representative values for $(s_A, s_B, s_C, s_{AB}, s_{BC})$. Note here that the discontinuities come from enforcing the consistency conditions.

plot3d(dedAC(0.1,0.1,0.4,sAB,sBC),sAB=0..1,sBC=0..1,numpoints=400, resolution = 400, axes=BOXED, labels=[sAB,sBC,dedAC]);

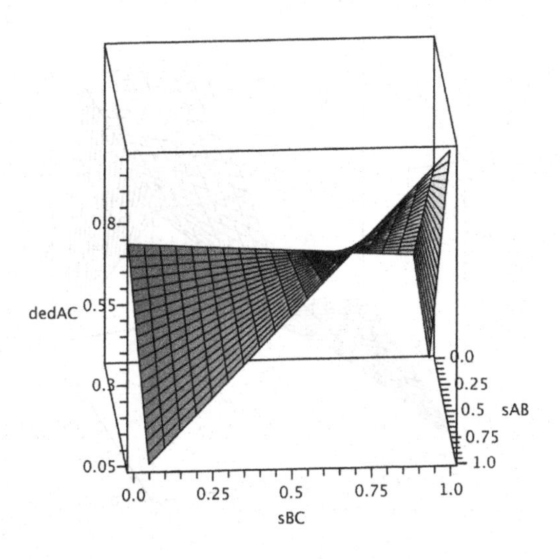

plot3d(dedAC(0.1,0.1,0.8,sAB,sBC),sAB=0..1,sBC=0..1,numpoints=400, resolution = 400, axes=BOXED, labels=[sAB,sBC,dedAC]);

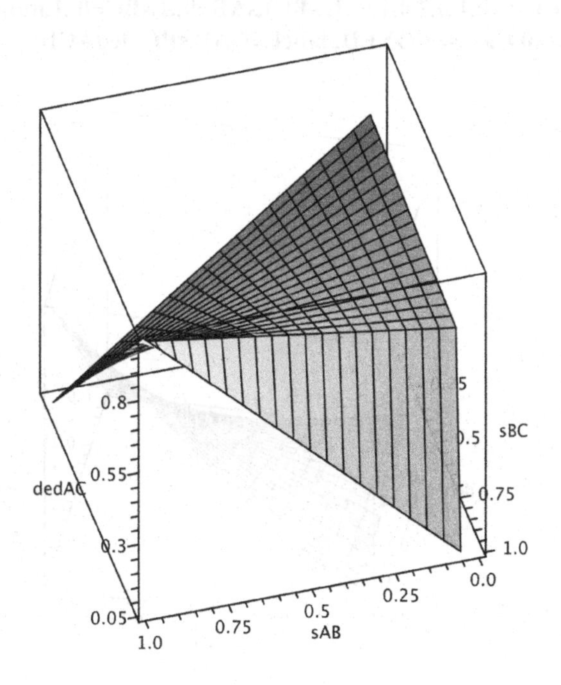

plot3d(dedAC(0.1,0.1,.95,sAB,sBC),sAB=0..1,sBC=0..1,numpoints=800, resolution = 400, axes=BOXED, labels=[sAB,sBC,dedAC]);

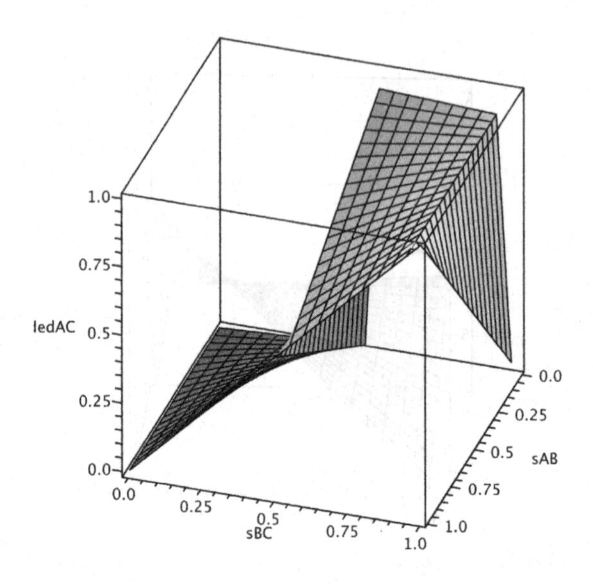

plot3d(dedAC(0.1,0.2,0.1,sAB,sBC),sAB=0..1,sBC=0..1,numpoints=800, resolution = 400, axes=BOXED, labels=[sAB,sBC,dedAC]);

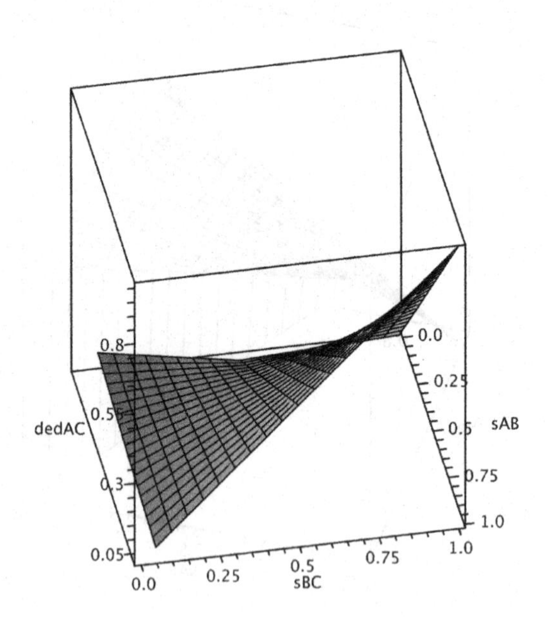

plot3d(dedAC(0.1,0.6,0.5,sAB,sBC),sAB=0..1,sBC=0..1,numpoints=800,axes=BOXED, labels=[sAB,sBC,dedAC]);

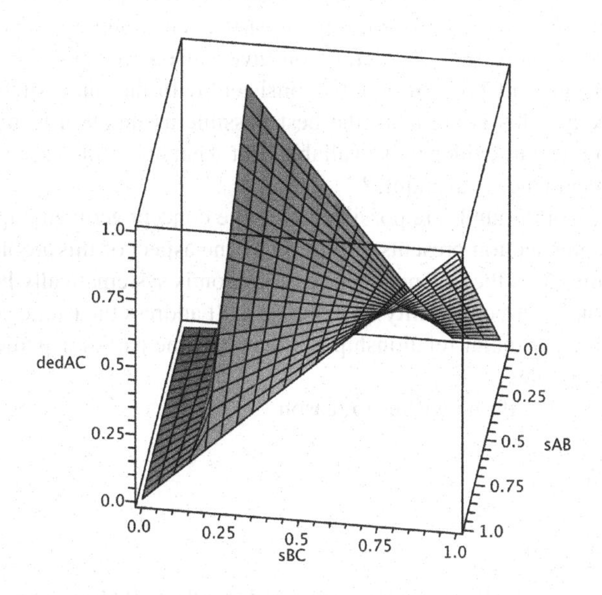

plot3d(dedAC(0.1,0.9,0.4,sAB,sBC),sAB=0..1,sBC=0..1,numpoints=800,labels=[sAB,sBC,sAC],axes=BOXED);

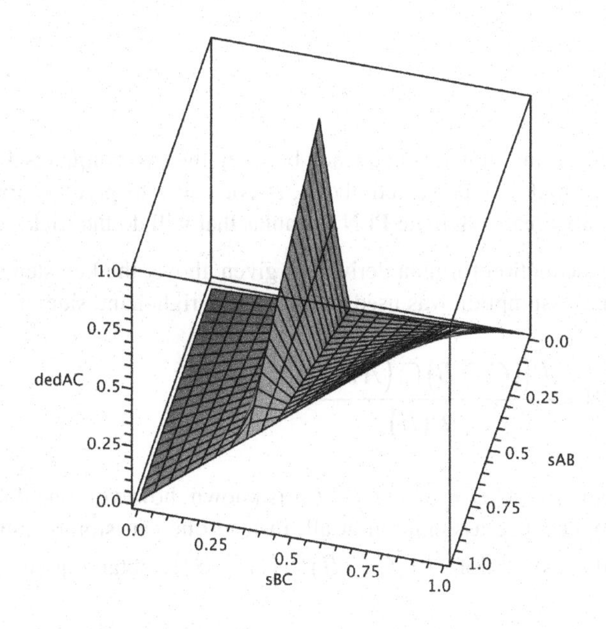

5.2.2.3 Deduction Accounting for Known Dependencies

Next, we present a minor variant of the above deduction strength formula. The heuristic inference formula derivation given earlier relies on an independence assumption, and the more rigorous derivation given just above assumes something similar, namely that all possible worlds consistent with the input strength values are equally likely. This is arguably the best possible assumption to make, in the case where no relevant evidence is available. But what about the case where additional relevant evidence *is* available?

A couple of workarounds are possible here. The concept geometry approach proposed in a later section presents a solution to one aspect of this problem – the fact that in some cases the independence assumption is systematically biased in a certain direction, relative to reality. But that doesn't address the (quite common) case where, for a particular relationship of interest, some partial dependency information is available.

For instance, what if one wishes to reason

```
Subset A B
Subset B C
|-
Subset A C
```

and one knows that $A \cap B$ and $C \cap B$ are *not* independent. Perhaps there exist conjunctions corresponding to these two compounds (these will be introduced more rigorously later on, but the basic idea should be clear), and perhaps there is an Equivalence or ExtensionalEquivalence relationship between these two Predicates, such as

```
ExtensionalEquivalence <.6>
    AND A B
    AND C B
```

indicating a significant known dependency between the two conjuncts. In this case the deduction formula, as presented above, is not going to give accurate results. But one can easily create alternate PLN formulas that will do the trick.

In the heuristic deduction formula derivation given above, the key step was where an independence assumption was used to split up the right-hand side of

$$P((C \cap A)|B) = \frac{P((C \cap B) \cap (A \cap B))}{P(B)}.$$

In the case where *(A AND B AND C)* <.6> is known, however, one doesn't need to use the independence assumption at all. Instead one can simply substitute the value .6 for the expression $P((C \cap B) \cap (A \cap B))$, obtaining the deduction formula

$$\frac{.6}{P(B)P(A)} + \frac{(1 - P(B|A))(P(C) - P(B)P(C|B))}{1 - P(B)}.$$

If we also had the knowledge

```
(A AND C AND (NOT B) ) <.3>
```

then we could simplify things yet further and use the formula

$$\frac{.6}{P(B)P(A)} + \frac{.3}{(1 - P(B))P(A)}.$$

The principle illustrated in these examples may be used more generally. If one is given explicit dependency information, then one can incorporate it in PLN via appropriately simplified inference formulae. As will be clear when we discuss alternate inference rules below, the same idea works for induction and abduction, as well as deduction.

5.3 The Optimal Universe Size Formula

Now we return to the question of the universe-size parameter U. It turns out that if one has information about a significant number of “triple intersection probabilities” in a context, one can calculate an “optimal universe size” for PLN deduction in that context.

The deduction rule estimates

$$P(C|A) = P(B|A)P(C|B) + P(\neg B|A|)P(C|\neg B)$$

where $n_A = |A|$, $n_{AC} = |A \cap C|$, etc., this means

$$\frac{n_{AC}}{n_A} = \frac{n_{AB}n_{BC}}{n_A n_B} + \frac{n_{A,\neg B}n_{B,\neg C}}{n_A n_{\neg B}}$$

The second term can be expanded

$$\frac{n_{A,\neg B}n_{\neg B,C}}{n_A n_{\neg B}} = \frac{(n_A - n_{AB})(n_C - n_{BC})}{n_A(n_U - n_B)}$$

where n_U is the size of the universe.

On the other hand, with triple intersections, one can calculate

$$n_{AC} = n_{ABC} + n_{A,\neg B,C}.$$

It's interesting to compare this with the formula

$$n_{AC} = \frac{n_{AB} n_{BC}}{n_B} + \frac{(n_A - n_{AB})(n_C - n_{BC})}{n_U - n_B}$$

derived from multiplying the deduction rule by n_A. Clearly the k'th term of each formula matches up to the k'th term of the other formula, conceptually.

Setting the second terms of the two equations equal to each other yields a formula for n_U, the universe size. We obtain

$$n_{A,\neg B,C} = \frac{(n_A - n_{AB})(n_C - n_{BC})}{n_U - n_B}$$

$$(n_{AC} - n_{ABC}) = \frac{(n_A - n_{AB})(n_C - n_{BC})}{n_U - n_B}$$

$$n_U = n_B + \frac{(n_A - n_{AB})(n_C - n_{BC})}{(n_{AC} - n_{ABC})}$$

For example, suppose $n_A=n_B=n_C=100$, $n_{AC} = n_{AB} = n_{BC} = 20$, $n_{ABC}= 10$ Then,

$$n_U = 100 + \frac{80 \cdot 80}{10} = 740$$

is the correct universe size.

The interesting thing about this formula is that none of the terms on the right side demand knowledge of the universe size – they're just counts, not probabilities. But the formula tells you the universe size that will cause the independence assumption in the second term of the deduction formula to come out exactly correctly!

Now, what use is this in practice? Suppose we know triple counts n_{ABC} for many triples (A,B,C), in a given context. Then we can take the correct universe sizes corresponding to these known triples and average them together, obtaining a good estimate of the correct universe size for the context as a whole.

We have tested this formula in several contexts. For instance, we analyzed a large corpus of newspaper articles and defined $P(W)$, where W is a word, as the percentage of sentences in the corpus that the word W occurs in at least once. The

Subset relationship strength from *W* to *V* is then defined as the probability that, if *V* occurs in a sentence, so does *W*; and the triple probability *P(W & V & X)* is defined as the percentage of sentences in which all three words occur. On this sort of data the universe size formula works quite well in practice. For instance, when the correct universe size – the number of sentences in the corpus – was around 650,000, the value predicted by the formula, based on just a few hundred triples, was off by less than 20,000. Naturally the accuracy can be increased by estimating more and more triples.

If one uses a concept geometry approach to modify the deduction rule, as will be discussed later, then the universe size calculation needs to be modified slightly, with a result that the optimal universe size increases.

5.4 Inversion, Induction, and Abduction

"Induction" and "abduction" are complex concepts that have been given various meanings by various different thinkers. In PLN we assign them meanings very close to their meanings in the NARS inference framework. Induction and abduction thus defined certainly do not encompass all aspects of induction and abduction as discussed in the philosophy and logic literature. However, from the fact that a certain aspect of commonsense induction or abduction is not encompassed completely by PLN induction or abduction, it should not be assumed that PLN itself cannot encompass this aspect. The matter is subtler than that.

For instance, consider abductive inference as commonly discussed in the context of scientific discovery: the conception of a theory encompassing a set of observations, or a set of more specialized theories. In a PLN-based reasoning system, we suggest, the process of scientific abduction would have to incorporate PLN abduction along with a lot of other PLN inference processes. Scientific abduction would be achieved by means of a non-inferential (or at least, not entirely inferential) process of hypothesis generation, combined with an inferential approach to hypothesis validation. The hypothesis validation aspect would not use the abduction rule alone: executing any complex real-world inference using PLN always involves the combination of a large number of different rules.

Given the above deduction formula, the definition of induction and abduction as defined in PLN is a simple matter. Each definition involves a combination of the deduction rule with a single use of the "inversion formula," which is simply Bayes' rule.

Inversion consists of the inference problem: Given $P(A)$, $P(B)$, and $P(A|B)$, find $P(B|A)$. The solution to this is:

$$P(B|A)=\frac{P(A|B)P(A)}{P(B)}$$

$$s_{AB}=\frac{s_{BA}s_A}{s_B}$$

which is a simple case of Bayes rule.

Induction consists of the inference problem: Given $P(A)$, $P(B)$, $P(C)$, $P(A|B)$ and $P(C|B)$, find $P(C|A)$. Applying inversion within the deduction formula, we obtain for this case:

$$P(C|A)=\frac{P(A|B)P(C|B)P(B)}{P(A)}+\left[1-\frac{P(A|B)P(B)}{P(A)}\right]\left[\frac{P(C)-P(B)P(C|B)}{1-P(B)}\right]$$

or

$$s_{AC}=\frac{s_{BA}s_{BC}s_B}{s_A}+\left(1-\frac{s_{BA}s_B}{s_A}\right)\left(\frac{s_C-s_Bs_{BC}}{1-s_B}\right)$$

In Maple notation, the induction formula

```
indAC := (sA, sB, sC, sBA, sBC) -> sBA * sBC * sB / sA + (1 - sBA * sB / sA) *(sC - sB * sBC)/(1-sB)*(Heaviside(sBA-max(((sA+sB-1)/sB),0))-Heaviside(sBA-min(1,(sA/sB))))*(Heaviside(sBC-max(((sB+sC-1)/sB),0))-Heaviside(sBC-min(1,(sC/sB))));
```

is depicted for selected input values in the following figures:

plot3d(indAC(0.1,0.1,0.1,sBA,sBC),sBA=0..1,sBC=0..1,numpoints=800, resolution = 400, labels=[sBA,sBC,indAC],axes=BOXED);

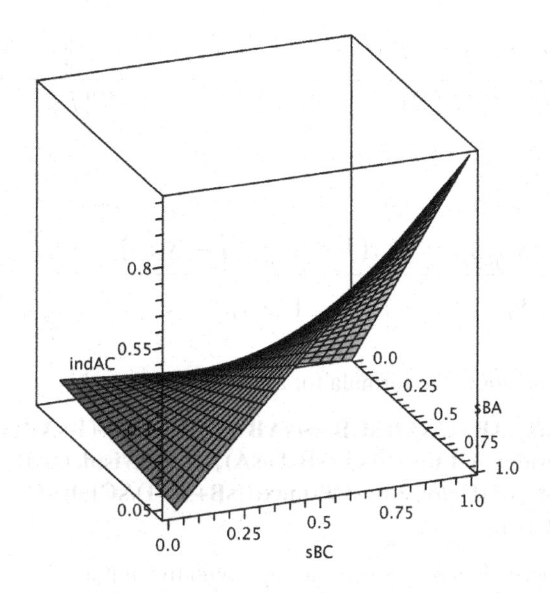

plot3d(indAC(0.4,0.1,0.1,sBA,sBC),sBA=0..1,sBC=0..1,numpoints=800, resolution = 400, labels=[sBA,sBC,indAC],axes=BOXED);

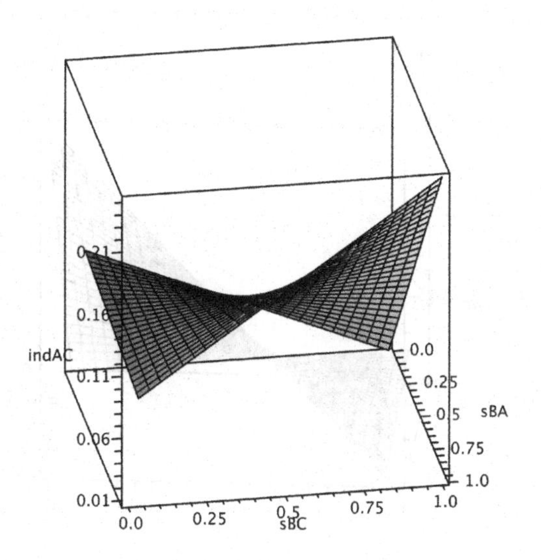

Next, abduction consists of the inference problem: Given *P*(*A*), *P*(*B*), *P*(*C*), *P*(*B*|*A*) and *P*(*B*|*C*), find *P*(*C*|*A*). In this case we need to turn *P*(*B*|*C*) into *P*(*C*|*B*), using inversion. We then obtain:

$$P(C|A) = \frac{P(B|A)P(B|C)P(C)}{P(B)} + \frac{P(C)[1-P(B|A)][1-P(B|C)]}{1-P(B)}$$

or

$$s_{AC} = \frac{s_{AB}s_{CB}s_C}{s_B} + \frac{s_C(1-s_{AB})(1-s_{CB})}{1-s_B}$$

In Maple notation, the formula for abduction is given by

abdAC:=(sA,sB,sC,sAB,sCB)->(sAB*sCB*sC/sB+(1-sAB)*(1-sCB)*sC/(1-sB))*(Heaviside(sAB-max(((sA+sB-1)/sA),0))-Heaviside(sAB-min(1,(sB/sA))))*(Heaviside(sCB-max(((sB+sC-1)/sC),0))-Heaviside(sCB-min(1,(sB/sC))));

We display here plots for two sets of representative inputs:
plot3d(abdAC(.1, .1, .1, sAB, sCB), sAB = 0 .. 1, sCB = 0 .. 1, numpoints = 800, resolution = 400, labels = [sAB, sCB, sAC], axes = BOXED);

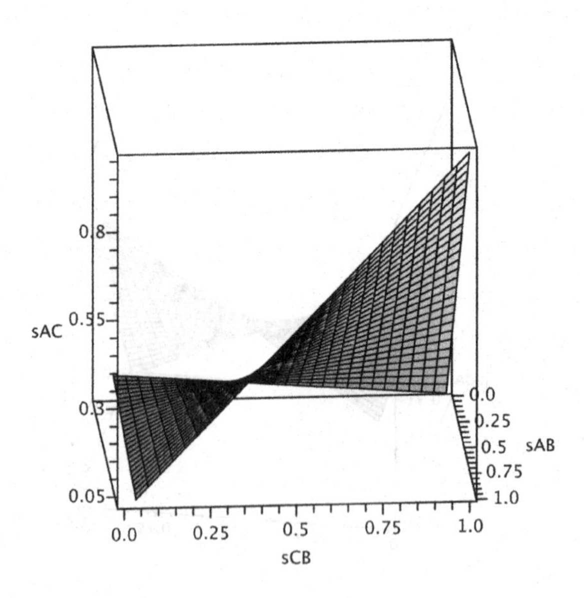

plot3d(abdAC(0.1,0.9,0.4,sAB,sCB),sAB=0..1,sCB=0..1,numpoints=800,labels =[sBA,sBC,sCA],axes=BOXED);

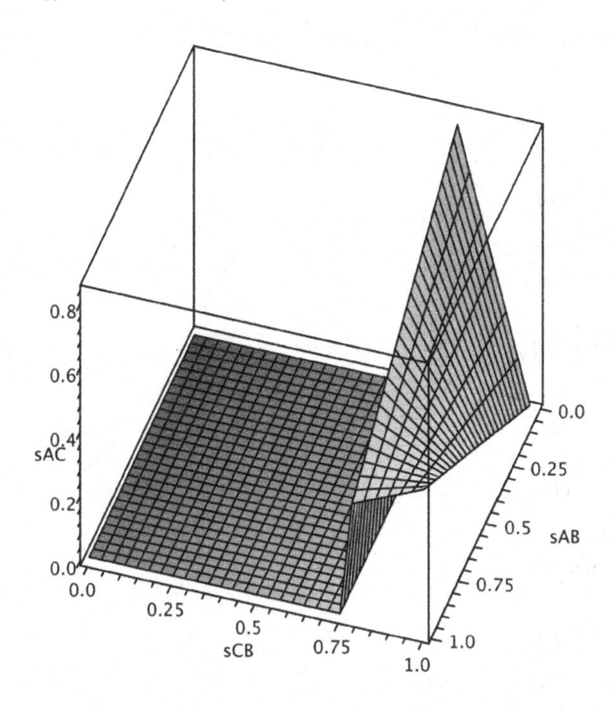

5.5 Similarity

Next we show how inference rules involving Similarity and associated symmetric logical relationships can be derived from the PLN rules for asymmetric logical relationships. The main rule here is one called "transitive similarity," which is stated as follows:

Transitive Similarity

Let sim_{AB}, sim_{BC}, and sim_{AC} represent the strengths of the Similarity relationships between A and B, between B and C, and between A and C, respectively. Given sim_{AB}, sim_{BC}, s_A, s_B, and s_C, the transitive similarity formula for calculating sim_{AC} then takes the form:

$$sim_{AC} = \frac{1}{\frac{1}{deduction(T_1,T_2)} + \frac{1}{deduction(T_3,T_4)} - 1}$$

where

$$T_1 = \frac{\left(1+\frac{s_B}{s_A}\right) sim_{AB}}{1+sim_{AB}}$$

$$T_2 = \frac{\left(1+\frac{s_C}{s_B}\right) sim_{BC}}{1+sim_{BC}}$$

$$T_3 = \frac{\left(1+\frac{s_B}{s_C}\right) sim_{BC}}{1+sim_{BC}}$$

$$T_4 = \frac{\left(1+\frac{s_A}{s_B}\right) sim_{AB}}{1+sim_{AB}}$$

and $deduction(T_1,T_2)$ and $deduction(T_3,T_4)$ are calculated using the independence-based deduction formula. That is, for example,

$$deduction(T_1,T_2) = T_1T_2 + (1-T_1)(s_C - s_BT_2)/(1-s_B)$$

The proof of the transitive similarity formula, to be given just below, contains the information needed to formulate several other rules for manipulating Similarity relationships, as well:

2inh2sim

Given s_{AC} and s_{CA}, estimate *simAC:*

$$sim_{AC} = \frac{1}{\frac{1}{s_{AC}} + \frac{1}{s_{CA}} - 1}$$

inh2sim

Given s_{AC} , s_A and s_C, estimate sim_{AC}:

$$sim_{AC} = \frac{1}{\frac{\left(1+\frac{s_A}{s_C}\right)}{s_{AC}}-1}$$

sim2inh

Given sim_{AB}, s_A and s_B, estimate s_{AB}

$$s_{AB} = \frac{\left(1+\frac{s_B}{s_A}\right)sim_{AB}}{1+sim_{AB}}$$

Proof of the Transitive Similarity Inference Formula

Assume one is given $P(A)$, $P(B)$, $P(C)$, and

$$sim_{AB} = \frac{|A\cap B|}{|A\cup B|} = \frac{P(A\cap B)}{P(A\cup B)}$$

$$sim_{BC} = \frac{|C\cap B|}{|C\cup B|} = \frac{P(C\cap B)}{P(C\cup B)}$$

and one wishes to derive

$$sim_{AC} = \frac{|A\cap C|}{|A\cup C|} = \frac{P(A\cap C)}{P(A\cup C)}$$

We may write

$$sim_{AC} = \frac{P(A \cap C)}{P(A \cup C)} = \frac{P(A \cap C)}{P(A) + P(C) - P(A \cap C)}$$

$$\frac{1}{sim_{AC}} = \frac{P(A) + P(C) - P(A \cap C)}{P(A \cap C)}$$

$$\frac{P(A)}{P(A \cap C)} + \frac{P(C)}{P(A \cap C)} - \frac{P(A \cap C)}{P(A \cap C)} = \frac{1}{P(C|A)} + \frac{1}{P(A|C)} - 1$$

So it suffices to estimate $P(C|A)$ and $P(A|C)$ from the given information. (The two cases are symmetrical, so solving one gives on the solution to the other via a simple substitution.)

Let's work on $P(C|A) = P(C \cap A)/P(A)$. To derive this, it suffices to estimate $P(B|A)$ and $P(C|B)$ from the given information, and then apply the deduction rule.

To estimate $P(B|A)$, look at

$$P(B|A) = \frac{|A \cap B|}{|A|} = sim_{AB} \frac{|A \cup B|}{|A|}$$

We thus need to estimate $|A \cup B|/|A|$, which can be done by

$$\frac{|A \cup B|}{|A|} = \frac{|A| + |B| - |A \cap B|}{|A|} = 1 + \frac{|B|}{|A|} - P(B|A).$$

We have the equation

$$P(B|A) = sim_{AB}\left(1 + \frac{|B|}{|A|} - P(B|A)\right)$$

$$P(B|A)(1 + sim_{AB}) = sim_{AB}\left(1 + \frac{|B|}{|A|}\right)$$

$$s_{AB} = P(B|A) = \frac{sim_{AB}\left(1 + \frac{|B|}{|A|}\right)}{(1 + sim_{AB})}.$$

A similar calculation gives us

$$s_{BC} = P(C|B) = \frac{sim_{BC}\left(1+\frac{|C|}{|B|}\right)}{(1+sim_{BC})}$$

with the deduction rule giving us a final answer via

$$s_{AC} = s_{AB}\, s_{BC} + (1-s_{AB})\,(s_C - s_B\, s_{BC})\,/\,(1-s_B).$$

Similarly, we may derive $P(A|C)$ from

$$s_{CA} = s_{CB}\, s_{BA} + (1-s_{CB})\,(s_A - s_B\, s_{BA})\,/\,(1-s_B)$$

where

$$s_{CB} = P(B|C) = \frac{\left(1+\frac{|B|}{|C|}\right)sim_{BC}}{1+sim_{BC}}$$

$$s_{BA} = P(A|B) = \frac{\left(1+\frac{|A|}{|B|}\right)sim_{AB}}{1+sim_{AB}}.$$

We thus obtain

$$sim_{AC} = \frac{1}{\frac{1}{s_{AC}}+\frac{1}{s_{CA}}-1}.$$

5.5.1 Graphical Depiction of Similarity Inference Formulas

Next we give a visual depiction of the transitive similarity inference, and related, formulas. The required formulas are:

2inh2sim := (sAC,sCA) -> 1/(1/sAC + 1/sCA - 1);

```
sim2inh := (sA,sB,simAB) -> (1 + sB/sA) * simAB / (1 + simAB)
*(Heaviside(simAB-max(((sA+sB-1)),0))-Heaviside(simAB-
min(sA/sB,(sB/sA))));

inh2sim := (sAC,sCA) -> 1/( 1/sAC + 1/sCA - 1);

simAC := (sA, sB, sC, simAB, simBC) -> inh2sim( sim2inh(sB,sC,simBC),
sim2inh(sA,sB,simAB));
```

where, as before, multiplication by the term

```
Heaviside(simBC-max(((sB+sC-1)),0))-Heaviside(simBC-
min(sB/sC,(sC/sB)))
```

represents a consistency condition on the three inputs s_B, s_C, and sim_{BC}.

The transitive similarity rule then looks like:

```
plot3d(simAC(.2, .2, .2, simAB, simBC), simAB = 0 .. 1, simBC = 0 .. 1, axes =
BOXED, labels =[simAB, simBC, simAC],numpoints=400, resolution=400);
```

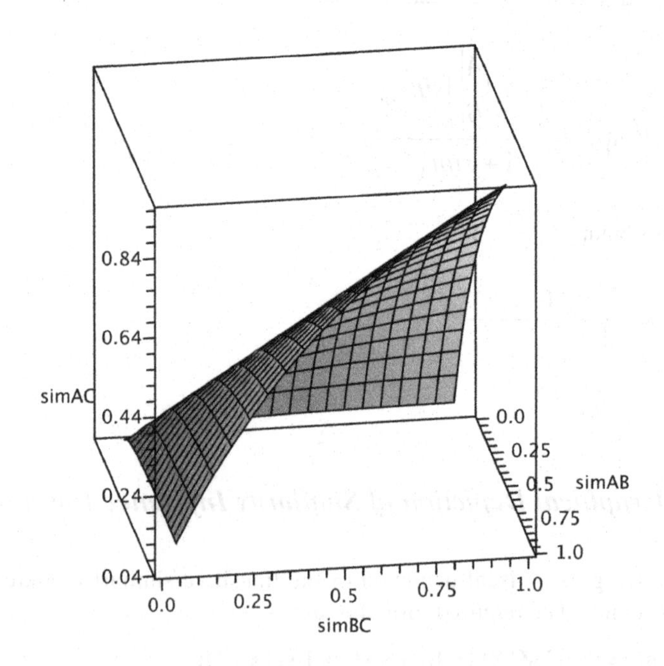

plot3d(simAC(.6, .2, .2, simAB, simBC), simAB = 0 .. 1, simBC = 0 .. 1, axes = BOXED, labels =[simAB, simBC, simAC],numpoints=400, resolution=400);

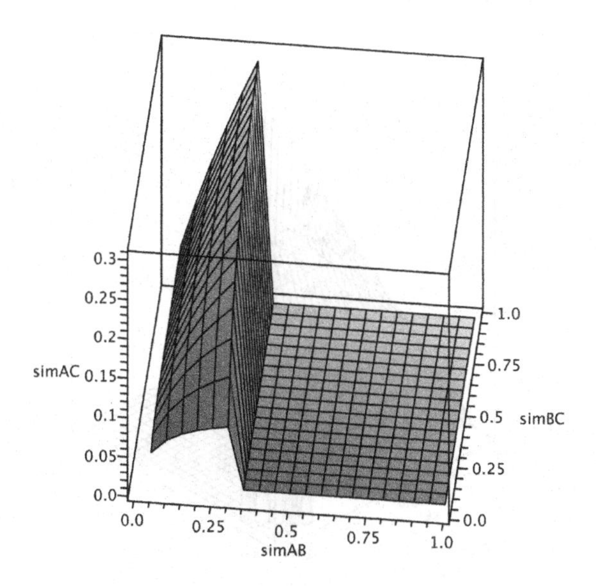

plot3d(simAC(.8, .1, .1, simAB, simBC), simAB =0 .. 1, simBC = 0 .. 1, axes = BOXED, labels =[simAB, simBC, simAC],numpoints=400, resolution=400);

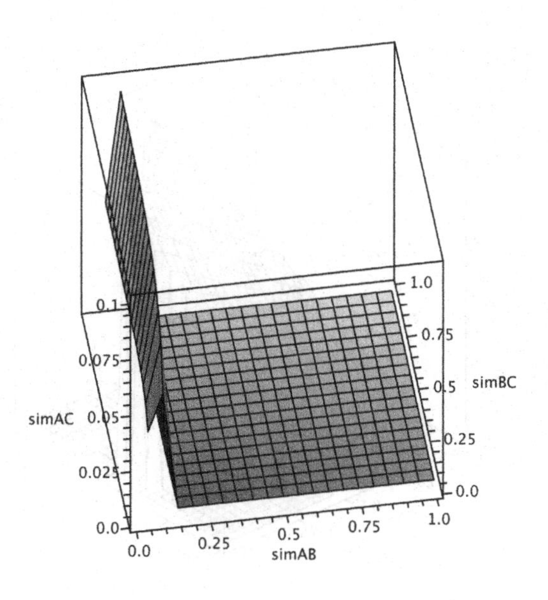

Next, a graphical depiction of the sim2inh rule is as follows:

plot3d(sim2inh(.2,sC,simBC),sC=0..1, simBC=0..1,axes=BOXED, numpoints=800, resolution =800);

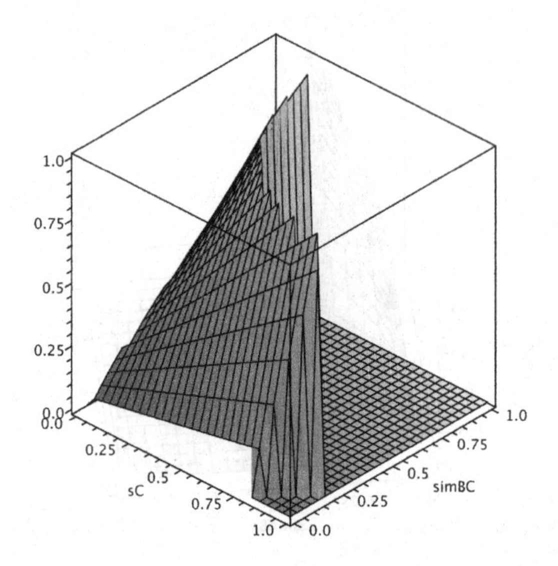

plot3d(sim2inh(.5,sC,simBC),sC=0.. 1,simBC=0.. 1, axes=BOXED, numpoints=800, resolution=800);

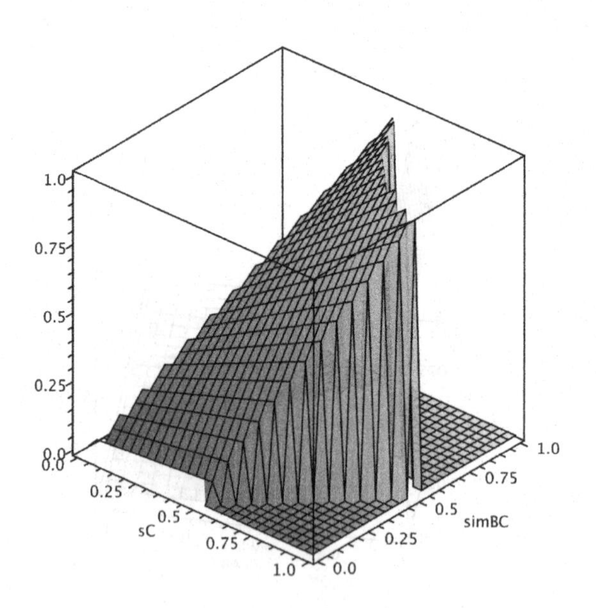

A visualization of the inverse, inh2sim rule is:

plot3d(inh2sim(sAC,sCA), sAC=0.01 ..1, sCA=0 ..1, axes=boxed);

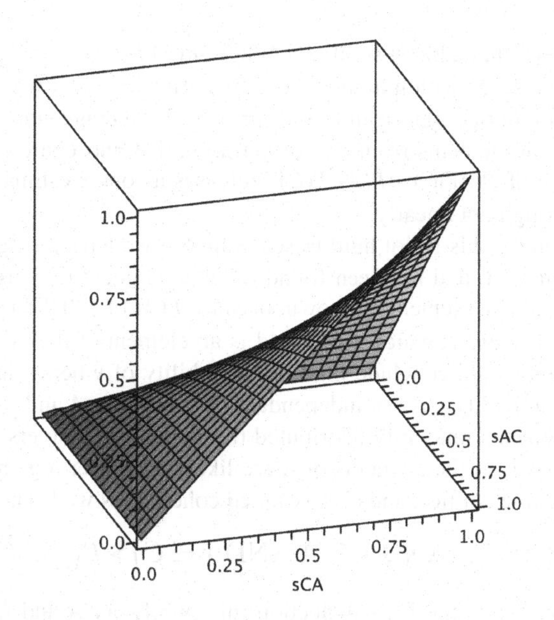

5.6 PLN Deduction via Concept Geometry

Now we return to the foundational issues raised in the derivation of the deduction formula above. Recall that the derivation of the deduction formula involved an independence assumption, which was motivated conceptually in terms of the process of "averaging over all possible worlds." The question that naturally arises is: how realistic is this process? The somewhat subtle answer is that the averaging-over-possible-worlds process seems the best approach (though certainly not perfectly accurate) in the case of deduction of s_{AC} with no prior knowledge of s_{AC}, but can be modified to good effect in the case where there is incomplete prior knowledge regarding the truth-value s_{AC}. In the case where there is incomplete prior knowledge of s_{AC}, the deduction formula may usefully be revised to take this knowledge into account.

The key observation here regards systematic violations of the independence assumption that occur due to the fact that most sets we reason on are "cohesive" – they consist of a collection of entities that are all fairly closely related to one another.

average, over all sets $\{A, C\}$, one will find $P(A\cap C) = P(A)\,P(C)$. However, if one has prior knowledge that

- $P(A\cap C) > 0$
- A and C are "cohesive" sets

then one knows that, almost surely, $P(A\cap C) > P(A)\,P(C)$.

A cohesive set S, within a universe U, is defined as a set whose elements have an average similarity significantly greater than the average similarity of randomly chosen pairs of elements from U. Furthermore, if A and C are cohesive, then the expected size of $P(A\cap C) - P(A)\,P(C)$ increases as one's estimate of $P(A\cap C)$ increases, although not linearly.

The reason for this is not hard to see. Suppose we have a single entity x which is an element of A, and it's been found to be an element of C as well. (The existence of this single element is known, because $P(A\cap C)>0$.) Then, suppose we're given another entity y, which we're told is an element of A. Treating y independently of x, we would conclude that the probability of y being an element of C is $P(C|A)$. But in reality y is not independent of x, because A and C are not sets consisting of elements randomly distributed throughout the universe. Rather, A is assumed cohesive, so the elements of A are likely to cluster together in the abstract space of possible entities; and C is assumed cohesive as well. Therefore,

$$P\left(y \in C \middle| y \in A \text{ AND } x \in A \text{ AND } x \in C\right) > P(y \in C)$$

In the context of the PLN deduction rule proof, our A and C are replaced by $A1 = A\cap B$ and $C1 = C\cap B$, but the basic argument remains the same. Note that if A and B are cohesive, then $A\cap B$ and $A\cap\neg B$ are also cohesive.

It's interesting to investigate exactly what the assumption of cohesion does to the deduction formula. Let us return to the heuristic derivation, in which we observed

$$P(C|A) = \frac{P(C\cap A)}{P(A)}$$

$$P(C\cap A) = P\big((C\cap B)\cap(A\cap B)\big) + P\big((C\cap\neg B)\cap(A\cap\neg B)\big)$$

If $A\cap B$, $C\cap B$, $A\cap\neg B$ and $C\cap\neg B$ are all cohesive, then we conclude that, on average, if

$$P\big((C\cap B)\cap(A\cap B)\big) > 0$$

then

$$P\big((C\cap B)\cap(A\cap B)\big) > P(C\cap B)P(A\cap B)$$

and, on average, if

$$P\left(\left(C \cap \neg B\right) \cap \left(A \cap \neg B\right)\right) > 0$$

then

$$P\left(\left(C \cap \neg B\right) \cap \left(A \cap \neg B\right)\right) > P\left(C \cap \neg B\right) P\left(A \cap \neg B\right)$$

If there is prior knowledge that $P(A \cap B \cap C)>0$ and/or $P(A \cap \neg B \cap C)>0$, then this can be used to predict that one or both of the terms of the deduction formula will be an underestimate.

On the other hand, what if there's prior knowledge about s_{AC}? If $s_{AC} > 0$ and A, B, and C are cohesive, this means that at least one of $P(A \cap B \cap C)>0$ or $P(A \cap \neg B \cap C)>0$ holds. And in most cases, unless B has a very special relationship with A and C or else s_{AC} is very small, both $P(A \cap B \cap C)>0$ and $P(A \cap \neg B \cap C)>0$ will hold. So given prior knowledge that $s_{AC} > 0$, we can generally assume that the deduction formula will be an underestimate.

In terms of the formal statement of the deduction formula, this implies that if we have prior knowledge that $s_{AC}>0$, then most of the time we'll have

$$E\left[f(x)\right] > s_{AB} s_{BC} + \frac{\left(1 - s_{AB}\right)\left(s_C - s_B s_{BC}\right)}{1 - s_B}$$

That is, it implies that the deduction formula, when applied to cohesive sets given prior knowledge that the value to be deduced is nonzero, will systematically underestimate strengths. Of course, this argument is just heuristic; to make it rigorous one would have to show that, given A, B, and C cohesive, the intersections used in this proof tend to display the probabilistic dependencies assumed. But these heuristic arguments seem very plausible, and they lead us to the following observation:

Observation: The Independence-Assumption-Based Deduction Formula Is Violated by the Assumption of Probabilistic Cohesion

Let U denote a set with $|U|$ elements. Let $Sub(m)$ denote the set of subsets of U containing m elements. Let s_A, s_B, s_C, s_{AB} and s_{BC} be numbers in [0,1], *all of which divide evenly into $|U|$. Let S be a set of cohesive subsets of U. Let*

$$f(x) = P\left[P\left(C|A\right) = x \,\middle|\, \begin{array}{l} A \in S \cap Sub\left(|U| s_A\right), B \in S \cap Sub\left(|U| s_B\right), \\ C \in S \cap Sub\left(|U| s_C\right), P\left(B|A\right) = s_{AB}, \\ P\left(C|B\right) = s_{BC}, P\left(C|A\right) > 0 \end{array} \right]$$

Then, where E() denotes the expected value (mean),

$$E\left[f(x)\right] > s_{AB}s_{BC} + \frac{(1-s_{AB})(s_C - s_B s_{BC})}{1-s_B}.$$

This observation is borne out by our practical experience applying PLN, both to simulated mathematical datasets and to real-world data.

So independence-assumption-based PLN deduction is flawed – what do we do about it? There is no simple and complete solution to the problem, but we have devised a heuristic approach that seems to be effective. We call this approach "concept geometry," and it's based on replacing the independence assumption in the above deduction rule with a different assumption regarding the "shapes" of the Nodes used in deduction. For starters we have worked with the heuristic assumption of "spherical" Nodes, which has led to some interesting and valuable corrections to the independence-assumption-based deduction formula. But the concept geometry approach is more general and can work with basically any assumptions about the "shapes" of Nodes, including assumptions according to which these "shapes" vary based on context.

5.6.1 Probabilistic Deduction Based on Concept Geometry

The concept-geometry approach begins with a familiar set-up:

$$P(C|A) = \frac{P(A \cap C)}{P(A)}$$

$$P(A \cap C) = P(A \cap C \cap B) + P(A \cap C \cap \neg B)$$

From this point on we proceed a little differently than in ordinary PLN deduction. If it is already believed that $P(A \cap C \cap B) > 0$, then one uses

$$P(A \cap C \cap B) = f(A \cap B, C \cap B)$$

where f is a function to be defined below. Otherwise, one uses

$$P(A \cap C \cap B) = P(A \cap B)P(C|B)$$

If it is already believed that $P(A \cap C \cap \neg B) > 0$, then one uses

$$P(A \cap C \cap \neg B) = f(A \cap \neg B, C \cap \neg B)$$

Otherwise, one uses

$$P(A \cap C \cap \neg B) = P(A \cap \neg B) P(C | \neg B)$$

One then computes $P(C|A)$ from $P(A|B)$, $P(C|B)$, $P(A|\neg B)$, $P(C|\neg B)$
These may be computed from the premise links using

$$P(A|B) = \frac{P(B|A)P(A)}{P(B)}$$

$$P(A|\neg B)P(\neg B) + P(A|B)P(B) = P(A)$$

so

$$P(A|\neg B) = \frac{\left[P(A) - P(A|B)P(B)\right]}{P(\neg B)} = \frac{P(A)\left[1 - P(B|A)\right]}{1 - P(B)}$$

$$P(C|\neg B) = \frac{\left[P(C) - P(C|B)P(B)\right]}{1 - P(B)}$$

The function $f(x,y)$ is estimated using a collection of model sets, according to the formula

`f(x,y) = expected (probability) measure of the intersection of a model set with measure x and a model set with measure y, assuming that the two model sets have nonzero intersection.`

The following theorem allows us to calculate $f(x, y)$ in some cases of interest. Specifically, suppose the model sets are "k-spheres" on the surface of a $(k+1)$-sphere. A special case of this is where the model sets are intervals on the boundary of the unit circle. Let $V(d)$ denote the volume of the k-sphere with diameter d, assuming a measure in which the volume of the whole surface of the $(k+1)$-sphere is 1. Given a measure x, let $diam(x)$ denote the diameter of the k-sphere with measure x.

Theorem 3: For model sets defined as k-spheres on the surface of a $(k+1)$-sphere, we have

$$f(x,y) = \frac{xy}{V(diam(x) + diam(y))}$$

Proof:

Let $r = diam(x)$, $q = diam(y)$; equivalently $x = V(r)$, $y = V(q)$. The idea is that the expected volume $f(x,y)$ can be expressed as

$$P\left(\|X - z\| \le r, \|Y - z\| \le q \,\middle|\, \|X - Y\| \le r + q\right)$$

Here X, Y are the centers of the spheres, z is a point belonging to the intersection of the spheres, and $\|V\|$ is the norm of vector V. We assume X, Y, z are chosen uniformly. The first two inequalities express the condition that z belongs to the intersection of spheres, the third inequality expresses the condition that these two spheres have a non-empty intersection. Using the definition of conditional probability, we have

$$\frac{P\left(\|X - z\| \le r, \|Y - z\| \le q, \|X - Y\| \le q + r\right)}{P\left(\|X - Y\| \le q + r\right)} = \frac{P\left(\|X - z\| \le r, \|Y - z\| \le q\right)}{P\left(\|X - Y\| \le q + r\right)} = \frac{V(r)V(q)}{V(q + r)}.$$

QED

As a special case, for intervals on a circle with circumference 1 we get

$$f(x,y) = \frac{xy}{\min(1, x + y)}$$

Note here a reasonably close parallel to the NARS deduction formula

$$\frac{xy}{x + y - xy}$$

For circles on the sphere, we obtain

$$f(x,y) = \frac{xy}{\min\left(1, x + y - 2xy + 2\sqrt{xy(1 - x)(1 - y)}\right)}$$

For the sake of concreteness, in the rest of this section we'll discuss mostly the interval formula, but in practice it is important to tune the dimensionality parameter to the specific domain or data source at hand, in order to minimize inference error. Potentially, if one has a number of different formulas from spheres in different spaces, and has some triple-intersection information from a dataset, then one can survey the various lookup tables and find the one most closely matching the dataset. To perform deduction based on these ideas, we need a way to estimate the odds that $P(A \cap B \cap C) > 0$. Most simply we could estimate this assuming knowledge that $P(A\cap C) > 0$, $P(B\cap C) > 0$, $P(A\cap B) > 0$ (and assuming we know that *A*, *B*, *C* are cohesive sets).

More sophisticatedly, we could use knowledge of the values of these binary intersection probabilities (and we will do so in the following, in a heuristic way). Let's use $G(A,B,C)$ to denote our estimate of the odds that $P(A\cap B\cap C) > 0$. In that case, if we model sets as intervals on the perimeter of the unit circle as suggested above, we have

$$P(A\cap B\cap C) = [1-G(A,B,C)]P(A\cap B)P(C|B) + G(A,B,C)P(A\cap B)P(C|B)\max\left(1,\frac{1}{P(A|B)+P(C|B)}\right)$$

$$= P(A\cap B)P(C|B)\left[1+G(A,B,C)\left(-1+\max\left(1,\frac{1}{P(A|B)+P(C|B)}\right)\right)\right]$$

$$= P(A\cap B)P(C|B)\left[1+G(A,B,C)\max\left(0,-1+\frac{1}{P(A|B)+P(C|B)}\right)\right]$$

$$= P(B|A)P(C|B)P(A)\left[1+G(A,B,C)\max\left(0,-1+\frac{1}{P(A|B)+P(C|B)}\right)\right]$$

So for the overall deduction rule we have

$$P(C|A) = P(B|A)P(C|B)\left[1+G(A,B,C)\max\left(0,-1+\frac{1}{P(A|B)+P(C|B)}\right)\right]$$
$$+ P(\neg B|A)P(C|\neg B)\left[1+G(A,\neg B,C)\max\left(0,-1+\frac{1}{P(A|\neg B)+P(C|\neg B)}\right)\right]$$

Now, to complete the picture, all we need is a formula for $G(A,B,C)$. One heuristic is to set $G(A,B,C) = 1$ if s_{AB}, s_{BC} and s_{AC} are all positive, and 0 otherwise, but obviously this is somewhat crude. One way to derive a heuristic for $G(A,B,C)$ is to assume that *A*, *B*, and *C* are 1-*D* intervals on the perimeter of the unit circle. In this case, as we'll show momentarily, we arrive at the formula

$$G(A,B,C) = P_{sym\_subsume}(A,B) + (1-P_{\text{sym\_subsume}}(A,B))P_{\text{int\_no\_subsume}}(A,B,C)$$

where the terms are defined as

$$P_{\text{sym\_subsume}}(A,B) = \frac{|P(A)-P(B)|}{P(A)+P(B)}$$

$$P_{\text{int\_no\_subsume}}(A,B,C)$$
$$= 1-\min[1+P(B)-P(A\cap B)-P(C), P(A)+P(B)-P(A\cap B)] + \max[P(B), 1-P(C)]$$

a fairly crude estimate, in that it uses knowledge of $P(A \cap B)$ and ignores knowledge of the other two binary intersections. A better estimate could be derived making use of the other two binary intersections as well; this estimate was chosen purely for simplicity of calculation. Also, one could make similar estimates using only $P(B \cap C)$ and only $P(A \cap C)$, and average these estimates together.

5.6.2 Calculation of a Heuristic Formula for G(A,B,C)

Here we will derive the above-mentioned heuristic formula for $G(A, B, C)$, defined as the odds that $P(A \cap B \cap C) > 0$. We do this by assuming that the sets *A*, *B*, and *C* are intervals on the boundary of the unit circle. Assume a measure that assigns the entire perimeter of the unit circle measure 1. Consider *B* as an interval extending from the top of the circle to the right. Next, consider *A*. If *A* totally subsumes *B* then obviously from $P(B \cap C) > 0$ we know $P(A \cap B \cap C) > 0$. The probability that *A* totally subsumes *B* is 0 if $P(A) < P(B)$. Otherwise, if we assume only the knowledge that $P(A \cap B) > 0$, the odds of *A* totally subsuming *B* is $\frac{P(A)-P(B)}{P(A)+P(B)}$.

If we assume knowledge of the value of $P(A \cap B)$, then a different and more complex formula would ensue. Overall we can estimate the probability that *A* totally subsumes *B* is

$$P_{\text{subsume}}(A,B) = \min\left[0, \frac{P(A)-P(B)}{P(A)+P(B)}\right]$$

Symmetrically, the probability that B totally subsumes A is $P_{\text{subsume}}(B,A)$ and we may define

$$P_{\text{sym\_subsume}}(A,B) = P_{\text{subsume}}(A,B) + P_{\text{subsume}}(B,A) = \frac{|P(A)-P(B)|}{P(A)+P(B)}$$

Suppose *A* does not totally subsume *B*, or vice versa? Without loss of generality we may assume that *A* overlaps only the right endpoint of *B*. Then let's compute the probability that *C* does not intersect $A \cap B$. For this to happen, the first (counting clockwise) endpoint of *C* [call this endpoint *x*] must occur in the portion of *A* that comes after *B*, a region with length $P(A)-P(A \cap B)$.

In order for *C* to intersect *B* but not $A \cap B$, we must have

$$1 < x + P(C) < 1 + P(B) - P(A \cap B)$$

i.e.,

$$1 - P(C) < x < 1 + P(B) - P(A \cap B) - P(C)$$

But by construction we have the constraint

$$P(B) < x < P(A) + P(B) - P(A \cap B)$$

So x can be anywhere in the interval

$$\left[\max(P(B), 1 - P(C)), \min(1 + P(B) - P(A \cap B) - P(C), P(A) + P(B) - P(A \cap B))\right]$$

So, the probability that C intersects $A \cap B$, given that neither A nor B totally subsumes each other, is

$$\begin{aligned} &P_{\text{int\_no\_subsume}}(A,B,C) \\ &= 1 - \min[1 + P(B) - P(A \cap B) - P(C), P(A) + P(B) - P(A \cap B)] + \max[P(B), 1 - P(C)] \end{aligned}$$

Our overall estimate is then

$$G(A,B,C) = P_{\text{sym\_subsume}}(A,B) + (1 - P_{\text{sym\_subsume}}(A,B))P_{\text{int\_no\_subsume}}(A,B,C)$$

5.7 Modus Ponens and Related Argument Forms

Next, we describe some simple heuristics via which PLN can handle certain inference forms that are common in classical logic, yet not particularly natural from a probabilistic inference perspective. We begin with the classic "modus ponens" (MP) inference formula, which looks like:

```
If P, then Q.
P.
Therefore, Q.
```

This is closely related to two other classic inference formulas: "modus tollens" (MT; proof by contrapositive):

```
If P, then Q.
Q is false.
Therefore, P is false.
```

and "disjunctive syllogism" (DS):

```
Either P or Q.
```

```
Not P.
Therefore, Q.
```

Of course, MT is immediately transformable into MP via

```
If ¬Q, then ¬P
¬Q
Therefore, ¬P
```

and DS is immediately transformable into MP via

```
If ¬P then Q.
¬P.
Therefore, Q.
```

We will also discuss a symmetrized version of modus ponens, of the form

```
If and only if P, then Q.
P.
Therefore, Q.
```

All these forms of inference are trivial when uncertainty is ignored, but become somewhat irritating when uncertainty is included. Term logic, in which the basic inference step is deduction from two inheritance relationships, plays much more nicely with uncertainty. However, it is still possible to handle MP and associated inferences within a probabilistic term logic framework, at the cost of introducing a simple but highly error-prone heuristic.

5.7.1 Modus Ponens

In PLN terms, what MP looks like is the inference problem

```
Inheritance A B <sAB>
A <sA>
|-
B <?>
```

This is naturally approached in a PLN-deduction-ish way via

$$P(B) = P(B|A)P(A) + P(B|\neg A)P(\neg A)$$

But given the evidence provided, we can't estimate all these terms, so we have no choice but to estimate $P(B|\neg A)$ in some very crude manner. One approach is to set $P(B|\neg A)$ equal to some "default term probability" parameter; call it c. Then we obtain the rule

$$s_B = s_{AB}s_A + c(1 - s_A)$$

For instance suppose we know

```
Inheritance smelly poisonous <.8>
smelly <.7>
```

Then in the proposed heuristic approach, we're estimating

```
P(poisonous) = .8*.7 + c*.3
```

where say c = .02 is a default term probability

Note that while this is a very bad estimate, it is clearly better than just setting

```
P(poisonous) = .02
```

or

```
P(poisonous) = .8*.7
```

which seem to be the only ready alternatives.

It's worth noting that the approach to MP proposed here is conceptually different from the one taken in NARS, wherein (where *A* and *B* are statements)

```
Implication A B <sAB>
A <sA>
```

is rewritten as

```
Implication A B <sAB>
Implication K A <sA>
```

where *K* is a "universal statement," and then handled by ordinary NARS deduction. This kind of approach doesn't seem to work for PLN. The NARS and PLN rules for MP have a similar character, in that they both consist of a correction to the product s_A s_{AB}. But in NARS it's the same sort of correction as is used for ordinary deduction, whereas in PLN it's a slightly different sort of correction.

5.7.2 Symmetric Modus Ponens

Now we turn to a related inference case, the symmetric modus ponens; i.e.,

```
SimilarityRelationship A B <simAB>
A <sA>
|-
B <?>
```

There is an additional subtlety here, as compared to standard MP, because to convert the similarity into an inheritance you have to assume some value for $P(B)$. If we take the above heuristic MP formula for s_B and combine it with the sim2inh rule

```
Given simAB, sA and sB, estimate sAB :
sAB = (1 + sB/sA) * simAB /( 1 + simAB )
```

then we obtain

$$s_B = s_A\left(1+\frac{s_B}{s_A}\right)\frac{sim_{AB}}{1+sim_{AB}} + c(1-s_A)$$

which (after a small amount of algebra) yields the heuristic formula

$$s_B = s_A sim_{AB} + c(1-s_A)(1+sim_{AB})$$

For example (to anticipate our later discussion of PLN-HOI), this is the rule we'll need to use for the higher-order inference

```
Equivalence <.8>
    Evaluation friendOf ($X,$Y)
    Evaluation friendOf ($Y,$X)
Evaluation friendOf(Ben, Saddam) <.6>
|-
Evaluation friendOf(Saddam, Ben) <?>
```

Given these numbers the answer comes out to .48 + .72*c. If we set c = .02 then we have .49 as an overall estimate.

5.8 Converting Between Inheritance and Member Relationship

Next we discuss conversion rules between Subset and Member relationships. The heuristic member-to-inheritance inference rule lets you reason, e.g.,

```
Member Ben Americans <tv1>
|-
Inheritance {Ben} Americans <tv2>
```

where {Ben} is the term with

```
Member Ben {Ben}
```

and no other members.

This conversion maps fuzzy truth-values into probabilistic truth-values. While the precise form of the member-to-inheritance rule depends upon the form truth-value being used, all forms follow the same fundamental idea: they all keep the mean strength constant and decrease the confidence in the result. For indefinite truth-values we accomplish this decrease in confidence by increasing the interval width by a constant multiplicative factor. The heuristic inheritance-to-member inference rule goes in the opposite direction and also keeps the strength constant and decreases confidence in a similar manner.

The semantics of these rules is a bit tricky – in essence, they are approximately correct only for terms that represent cohesive concepts; i.e., for sets whose members share a lot of properties. They are badly wrong for Atoms representing random sets. They are also bad for sets like "Things owned by Izabela" which are heterogeneous collections, not sharing so many common properties. This issue is a subtle one and relates to the notion of concept geometry, to be discussed in a later section.

5.9 Term Probability Inference

The inference formulas given above all deal with inference on relationships, but they involve term probabilities as well as relationship probabilities. What about inference on term probabilities? Some kinds of Term-based inference require higher-order reasoning rules. For instance, if a PLN based reasoning system knows $s_{man} = s_{woman} = .5$, and it knows that

```
Inheritance human ( man OR woman) <.9999>
```

then it should be able to infer that s_{human} should not be .001. In fact it may well come to the wrong conclusion, that $s_{human} = .5$, unless it is given additional information such as

```
man AND woman <.00001>
```

in which case after applying some logical transformation rules it can come to the correct conclusion that $s_{human} = .9999$. This will become clearer when we discuss higher-order inference in the following chapter.

But it is valuable to explicitly encode term probability inference beyond that which comes indirectly via HOI. For instance, suppose the system knows that

```
Inheritance cat animal
```

and it sees evidence of a cat in its perceptual reality, thus increasing its probability estimate s_{cat}. Then, it should be able to adjust its probability estimate of s_{animal} indirectly as well, by inference. The rule here is a very simple one – just a rearrangement of Bayes' rule in a different order than is used for inversion. From

$$P(B) = \frac{P(B|A)P(A)}{P(A|B)}$$

one derives the formula

$$s_B = \frac{s_A s_{AB}}{s_{BA}}$$

or in other words

```
A <sA>
Inheritance A B <sAB>
|-
B <sB>
```

This brings us to a point to be addressed a few sections down: how to deal with circular inference. We will introduce a mechanism called an "inference trail," which consists in essence of a series of Atoms, so that the truth-value of each one is considered as partially determined by the truth-values of its predecessors in the series. If one is using trails, then term probability inference must be included in inference trails just as relationship probability inference is, so that, after inferring s_{animal} from s_{cat} and $s_{cat,animal}$, one does not then turn around and infer $s_{cat,\ animal}$ from s_{animal} and s_{cat}.

In general, various mechanisms to be introduced in the following sections and discussed primarily in the context of relationship strengths may be assumed to apply in the context of term strengths as well. For example, the revision rule discussed in the following section works on term truth-values just as on relationship truth-values, and distributional truth-values as discussed in Chapter 7 carry over naturally as well.

5.10 Revision

In this section we introduce another important PLN inference rule: the "revision" rule, used to combine different estimates of the truth-value of the same Atom. In general, truth-value estimates to be merged via revision may come from three different sources:

- External sources, which may provide different estimates of the truth-value of the same Atom (e.g., the New York Times says a certain politician is very trustworthy, but the Wall Street Journal says he is only moderately trustworthy)
- Various cognitive processes, inside an integrative AI system, provide different estimates of the truth-value of the same Atom

- Different pathways of reasoning, taken internally by PLN, which may provide different estimates of the truth-value of the same Atom

Generally speaking, it is the latter source which provides the greatest number of truth-value estimates, and which makes the revision formula an extremely critical part of PLN.

Pei Wang, in (Wang, 1993), has given an excellent review of prior approaches to belief revision – including probabilistic approaches – and we mostly agree with his assessments of their shortcomings. His own approach to revision involves weighted averaging using a variable called "confidence," which is quite similar to PLN's "weight of evidence" – and although we don't think it's ideal (otherwise the detailed developments of this section wouldn't be necessary), we do find his approach basically sensible, unlike most of the approaches in the literature. The key point that Wang saw but most other belief revision theorists did not is that, in order to do revision sensibly, truth-values need at least two components, such as exist in NARS and in PLN but not in most uncertain inference frameworks.

5.10.1 Revision and Time

Before proceeding with the mathematics, it's worth mentioning one conceptual issue that arises in the context of revision: the nonstationary nature of knowledge. If different estimates of the truth-value of an Atom were obtained at different times, they may have different degrees of validity on this account. Generally speaking, more weight needs to be given to the relationship pertaining to more recent evidence – but there will of course be many exceptions to this rule.

In an integrative AI context, one approach to this problem is to push it out of the domain of logic and into the domain of general cognition. The treatment of revision in PLN reflects our feeling that this is likely the best approach. We assume, in the PLN revision rule, that all information fed into the reasoning system has already been adapted for current relevance by other mental processes. In this case all information can be used with equal value. In an integrative AI system that incorporates PLN, there should generally be separate, noninferential processes that deal with temporal discounting of information. These processes should work, roughly speaking, by discounting the count of Atoms' truth-values based on estimated obsoleteness, so that older estimates of the same Atom's truth-value will, all else being equal, come to have lower counts. According to the revision formulas to be given below, estimates with lower counts will tend to be counted less in the revision process.

However, because in any complex AI system, many cognitive processes are highly nondeterministic, it cannot be known for sure that temporal updating has been done successfully on two potential premises for revision. Hence, if the premises are sufficiently important and their truth-values are sufficiently different, it may be appropriate to specifically request a temporal updating of the premises, and carry through with the revision only after this has taken place. This strategy can also be used with other inference rules as well: requesting an update of the premises before deduction, induction, or abduction takes place.

5.10.2 A Heuristic Revision Rule for Simple Truth-values

In this section we give a simple, heuristic revision rule for simple truth-values consisting of (strength, weight of evidence) pairs. In the following chapter we present a subtler approach involving indefinite probabilities; and in following subsections of this section we present some subtler approaches involving simple truth-values. All in all, belief revision is a very deep area of study, and the PLN framework supports multiple approaches with differing strengths and weaknesses.

Suppose s_1 is the strength value for premise 1 and s_2 is the strength value for premise 2. Suppose further that n_1 is the count for premise 1 and n_2 is the count for premise 2. Let $w_1 = n_1/(n_1+n_2)$ and $w_2 = n_2/(n_1+n_2)$ and then form the conclusion strength value $s = w_1 s_1 + w_2 s_2$.

In other words, a first approximation at a reasonable revision truth-value formula is simply a weighted average. We may also heuristically form a count rule such as $n = n_1 + n_2 - c \min(n_1, n_2)$, where the parameter c indicates an assumption about the level of interdependency between the bases of information corresponding to the two premises. The value c=1 denotes the assumption that the two sets of evidence are completely redundant; the value c=0 denotes the assumption that the two sets of evidence are totally distinct. Intermediate values denote intermediate assumptions.

5.10.3 A Revision Rule for Simple Truth-values Assuming Extensive Prior Knowledge

In this and the next few subsections, we dig more deeply into the logic and arithmetic of revision with simple truth-values. This material is not currently utilized in our practical PLN implementation, nor has it been fully integrated with the indefinite probabilities approach. However, we feel it is conceptually valuable, and represents an important direction for future research.

To illustrate conceptually the nature of the revision process, we consider here the simple case of an extensional inheritance relation

```
L = Subset A B
```

Suppose we have two different estimates of this relationship's strength, as above. We consider first the case where the weight of evidence is equal for the two different estimates, and then take a look at the role of weight of evidence in revision.

Suppose that the two different estimates were obtained by looking at different subsets of *A* and *B*. Specifically, suppose that the first estimate was made by looking at elements of the set *C*, whereas the second estimate was made by looking at elements of the set *D*.

Then we have

$$s_1 = \frac{|A \cap B \cap C|}{|A \cap C|}$$

$$s_2 = \frac{|A \cap B \cap D|}{|A \cap D|}$$

By revising the two strengths, we aim to get something like

$$s_3^* = \frac{|A \cap B \cap (C \cup D)|}{|A \cap (C \cup D)|}$$

which is the inheritance value computed in the combined domain $C \cup D$.

Now, there is no way to exactly compute s_3* from s_1 and s_2, but we can compute some approximation $s_3 \cong s_3$*. Let's write

$$s_1 = \frac{x_1}{y_1}$$

$$s_2 = \frac{x_2}{y_2}$$

If we assume that *C* and *D* are independent in both *A* and $A \cap B$, then after a bit of algebra we can obtain the heuristic revision rule:

$$s_3 = \frac{x_1 + x_2 - \dfrac{x_1 x_2}{u}}{y_1 + y_2 - \dfrac{y_1 y_2}{u}}$$

The derivation goes as follows. Firstly, we have

$$|A \cap B \cap (C \cup D)| = |A \cap B \cap C| + |A \cap B \cap D| - |A \cap B \cap C \cap D|$$

$$|A \cap (C \cup D)| = |A \cap C| + |A \cap D| - |A \cap C \cap D|$$

We thus have

$$s_3^* = \frac{x_1 + x_2 - |A \cap B \cap C \cap D|}{y_1 + y_2 - |A \cap C \cap D|}$$

And we now have an opportunity to make an independence assumption. If we assume that *C* and *D* are independent within *A*, then we have

$$\frac{|A \cap C \cap D|}{|U|} = \left(\frac{|A \cap C|}{|U|}\right)\left(\frac{|A \cap D|}{|U|}\right)$$

(where *U* is the universal set) and thus

$$|A \cap C \cap D| = \frac{|A \cap C||A \cap D|}{|U|} = \frac{y_1 y_2}{u}$$

(where $u = |U|$). Similarly if we assume that *C* and *D* are independent within $A \cap B$, we have

$$|A \cap B \cap C \cap D| = \frac{|A \cap B \cap C||A \cap B \cap D|}{|U|} = \frac{x_1 x_2}{u}$$

which yields the revision rule stated above:

$$s_3 = \frac{x_1 + x_2 - \frac{x_1 x_2}{u}}{y_1 + y_2 - \frac{y_1 y_2}{u}}$$

As with the deduction rule, if the system has knowledge about the intersections $|A \cap C \cap D|$ or $|A \cap B \cap C \cap D|$, then it doesn't need to use the independence assumptions; it can substitute its actual knowledge.

Now, one apparent problem with this rule is that it doesn't actually take the form

$$s_3 = f(s_1, s_2)$$

It requires a little extra information. For instance suppose we know the strength of

```
Subset A C <sAC>
```

Then we know

$$y_1 = |A \cap C| = s_{AC} P(A) u$$

$$x_1 = s_1 y_1$$

so we can compute the needed quantities (x_1, y_1) from the given strength s_1 plus the extra relationship strength s_{AC}.

5.10.4 *A Revision Rule for Simple Truth-values Assuming Extensive Prior Knowledge*

But – following up the example of the previous subsection – what do we do in the case where we don't have any extra information, just the premise strengths s_1 and s_2? Suppose we assume y_1=y_2=q, for simplicity, assuming no knowledge about either. Then s_1=x_1/q, s_2 = x_2/q , and we can derive a meaningful heuristic revision formula. In this case the

$$s_3 = \frac{x_1 + x_2 - \frac{x_1 x_2}{u}}{y_1 + y_2 - \frac{y_1 y_2}{u}}$$

given above specializes to

$$s_3 = \frac{s_1 q + s_2 q - \frac{s_1 s_2 q q}{u}}{q + q - \frac{q q}{u}} = \frac{s_1 + s_2 - \frac{s_1 s_2 q}{u}}{2 - \frac{q}{u}}$$

which can be more simply written (setting c=q/u)

$$s_3 = \frac{s_1 + s_2 - s_1 s_2 c}{2 - c}$$

Here c is a critical parameter, and as the derivation shows, intuitively

- $c = 1$ means that both premises were based on the whole universe (which is generally implausible, because if this were the case they should have yielded identical strengths)
- $c = .5$ means that the two premises were each based on half of the universe
- $c = 0$ means that the premises were each based on negligibly small percentages of the universe

Different c values may be useful in different domains, and it may be possible in some domains to tune the c value to give optimal results. Concretely, if $c = 0$, then we have

$$s_3 = \frac{s_1 + s_2}{2}$$

i.e., the arithmetic average. We have derived the intuitively obvious "averaging" heuristic for revision, based on particular assumptions about the evidence sets under analysis.

But suppose we decide c=.5, to take another extreme. Then we have

$$s_3 = \frac{2}{3}\left(s_1 + s_2 - .5 s_1 s_2\right)$$

Suppose $s_1 = s_2 = .1$, then the average formula gives s_3=.1, but this alternate formula gives

$$s_3 = \frac{2}{3}(0.1 + 0.1 - 0.5 \cdot 0.1 \cdot 0.1) = 0.13$$

which is higher than the average!

So what we find is that the averaging revision formula effectively corresponds to the assumption of an infinite universe size. With a smaller universe, one obtains revisions that are higher than the averaging formula would yield. Now, for this to be a significant effect in the course of a single revision one has to assume the universe is very small indeed. But iteratively, over the course of many revisions, it's possible for the effect of a large but non-infinite universe to have a significant impact.

In fact, as it turns out, some tuning of c is nearly always necessary – because, using the above heuristic formula, with c nonzero, one obtains a situation in which revision gradually increases relationship strengths. Applied iteratively for long enough, this may result in relationship strengths slowly drifting up to 1. And this may occur even when not correct – because of the independence assumption embodied in the $(s_1\ s_2)$ term of the rule. So if c is to be set greater than zero, it must be dynamically adapted in such a way as to yield "plausible" results, a notion which is highly domain-dependent.

5.10.5 *Incorporating Weight of Evidence into Revision of Simple Truth-values*

Next, one can modify these rules to take weight of evidence into account. If we define

$$v_1 = \frac{2d_1}{d_1 + d_2}$$
$$v_2 = 2 - v_1$$

then some algebra yields

$$s_3 = \frac{v_1 x_1 + v_2 x_2 - \frac{x_1 x_2}{u}}{v_1 y_1 + v_2 y_2 - \frac{y_1 y_2}{u}}$$

which yields the corresponding heuristic formula

$$s_3 = \frac{v_1 s_1 + v_2 s_2 - s_1 s_2 c}{2 - c}$$

The algebra underlying this involved going back to the original derivation of the revision rule, above. Where we said

$$|A \cap B \cap (C \cup D)| = |A \cap B \cap C| + |A \cap B \cap D| - |A \cap B \cap C \cap D|$$

$$|A \cap (C \cup D)| = |A \cap C| + |A \cap D| - |A \cap C \cap D|$$

we now instead have to say

$$|A \cap B \cap (C \cup D)| = v_1 |A \cap B \cap C| + v_2 |A \cap B \cap D| - |A \cap B \cap C \cap D|$$

$$|A \cap (C \cup D)| = v_1|A \cap C| + v_2|A \cap D| - |A \cap C \cap D|$$

where

$$v_1 = \frac{2d_1}{d_1 + d_2}$$
$$v_2 = 2 - v_1$$

which is equivalent to the formulas given above.

5.10.6 Truth-value Arithmetic

As a kind of footnote to the discussion of revision, it is interesting to observe that one may define a kind of arithmetic on PLN two-component truth-values, in which averaging revision plays the role of the addition on strengths, whereas simple summation plays the role of revision on counts. We have not found practical use for this arithmetic yet, but such uses may emerge as PLN theory develops further.

The first step toward formulating this sort of truth-value arithmetic is to slightly extend the range of the set of count values. Simple two-component PLN truth-values, defined above, are values lying in $[0,1]\times[0, \infty]$. However, we can think of them as lying in $[0,1]\times[-\infty, \infty]$, if we generalize the interpretation of "evidence" a little bit, and introduce the notion of negative total evidence.

Negative total evidence is not so counterintuitive, if one thinks about it the right way. One can find out that a piece of evidence one previously received and accepted was actually fallacious. Suppose that a proposition P in the knowledge base of a system S has the truth-value (1,10). Then, for the system's truth-value for P to be adjusted to (.5,2) later on, it will have to encounter 9 pieces of negative evidence and 1 piece of positive evidence. That is, all 10 of the pre-July-19 observations must be proved fraudulent, and one new observation must be made (and this new observation must be negative evidence). So, we may say (anticipating the truth-value arithmetic to be introduced below):

$$(1,10) + (1,-9) + (0,1) = (0.5,2)$$

Now, suppose we have two truth-values T and S. We then define

$$T + S = W\left(W^{-1}(T) + W^{-1}(S)\right)$$

where the function W is a nonlinear mapping from the real plane into the region $[0,1] \times [-\infty, \infty]$.

Similarly, where c is a real number, we define the scalar multiple of a truth-value T by

$$c * T = W\left(W^{-1}(c) * W^{-1}(T)\right)$$

It is easy to verify that the standard vector space axioms (Poole, 2006) hold for these addition and scalar multiplication operations.

A vector space over the field F is a set V together with two binary operations,

- vector addition: $V \times V \rightarrow V$ denoted v + w, where v, w $\in$ V, and
- scalar multiplication: $F \times V \rightarrow V$ denoted a v, where a $\in$ F and v $\in$ V,

satisfying the axioms below. Four require vector addition to be an Abelian group, and two are distributive laws.

1. Vector addition is associative: For all u, v, w $\in$ V, we have u + (v + w) = (u + v) + w.
2. Vector addition is commutative: For all v, w $\in$ V, we have v + w = w + v.
3. Vector addition has an identity element: There exists an element 0 $\in$ V, called the zero vector, such that v + 0 = v for all v $\in$ V.
4. Vector addition has an inverse element: For all v $\in$ V, there exists an element w $\in$ V, called the additive inverse of v, such that v + w = 0.
5. Distributivity holds for scalar multiplication over vector addition: For all a $\in$ F and v, w $\in$ V, we have a (v + w) = a v + a w.
6. Distributivity holds for scalar multiplication over field addition: For all a, b $\in$ F and v $\in$ V, we have (a + b) v = a v + b v.
7. Scalar multiplication is compatible with multiplication in the field of scalars: For all a, b $\in$ F and v $\in$ V, we have a (b v) = (ab) v.
8. Scalar multiplication has an identity element: For all v $\in$ V, we have 1 v = v, where 1 denotes the multiplicative identity in F.

The zero of the truth-value + operator is *W*(0); the unit of the truth-value * operator is *W*(1). Note that axioms 4 and 6, which deal with closure, hold only because we have decided to admit negative amounts of total evidence.

The next question is, what is a natural nonlinear mapping function *W* to use here? What we would like to have is

$$W^{-1}\left(S_f(g), E^*(g)\right) = W^{-1}\left(\frac{E_f^+(g)}{E_f^+(g) + E_f^-(g)}, E_f^+(g) + E_f^-(g)\right)$$
$$= \left(E_f^+(g) - E_f^-(g), E_f^+(g) + E_f^-(g)\right)$$

In other words, we want

$$W(a-b, a+b) = \left(\frac{a}{a+b}, a+b\right).$$

This is achieved by defining

$$W(x,y) = \left(\frac{1+\frac{x}{y}}{2}, y\right)$$

$$W^{-1}(x,y) = (y(2x-1), y)$$

Now, finally getting back to the main point, the relationship between truth-value addition as just defined and PLN revision is not hard to see. We now lapse into standard PLN notation, with s for strength and N for count. Where $s_i = N^+_i/N_i$, we have

$$\begin{aligned}
&(s_1, N_1) + (s_2, N_2) = W\left(W^{-1}(s_1, N_1) + W^{-1}(s_2, N_2)\right) \\
&= W\left(N_1^+ - N_1^-, N_1\right) + \left(N_1^+ - N_2^-, N_1 + N_2\right) \\
&= W\left(\left(N_1^+ + N_2^+\right) - \left(N_1^- + N_2^-\right), \left(N_1^+ + N_2^+\right) + \left(N_1^- + N_2^-\right)\right) \\
&= \left(\frac{N_1^+ + N_2^+}{\left(N_1^+ + N_2^+\right) + \left(N_1^- + N_2^-\right)}, \left(N_1^+ + N_2^+\right) + \left(N_1^- + N_2^-\right)\right) \\
&= \left(w_1 s_1 + w_2 s_2, N_1 + N_2\right)
\end{aligned}$$

where

$$w_1 = \frac{N_1}{N_1 + N_2}$$

$$w_2 = \frac{N_2}{N_1 + N_2}$$

This is just the standard PLN revision rule: weighted averaging based on count. So we see that our truth-value addition rule is nothing but revision!

On the other hand, our scalar multiplication rule boils down to

$$c*(s,N) = c*\left(\frac{N^+}{N^+ + N^-}, N^+ + N^-\right)$$
$$= W\left(c*\left(N^+ - N^-\right), c*\left(N^+ + N^-\right)\right)$$
$$= \left(\frac{N^+}{N^+ + N^-}, c*\left(N^+ + N^-\right)\right)$$
$$= (s, c*N)$$

In other words, our scalar multiplication leaves strength untouched and multiplies only the total evidence count. This scalar multiplication goes naturally with revision-as-addition, and yields a two-dimensional vector space structure on truth-values.

5.11 Inference Trails

One issue that arises immediately when one implements a practical PLN system is the problem of circular inference (and, you may recall that this problem was briefly noted above in the context of term probability inference). For example, suppose one has

```
P(American) = <[0.05, 0.15], 0.9, 10>
P(silly) = <[0.3, 0.5], 0.9, 10>
P(human) = <[0.4, 0.6], 0.9, 10>
```

Then one can reason

```
Subset American human <[0.7, 0.9], 0.9, 10>
Subset human silly <[0.4, 0.45], 0.9, 10>
|-
Subset American silly <[0.370563, 0.455378], 0.9, 10>
```

but then one can reason

```
Subset human silly <[0.4, 0.45], 0.9, 10>
|-
Subset silly human <[0.377984, 0.595297], 0.9, 10>
```

and

```
Subset American silly <[0.370563, 0.455378], 0.9, 10>
Subset silly human <[0.377984, 0.595297], 0.9, 10>
```

```
|-
Subset American human <[0.408808, 0.578298], 0.9, 10>
```

which upon revision with the prior value for (Subset American human) yields

```
Subset American human <[0.578047, 0.705526], 0.9, 10>
```

Repeating this iteration multiple times, one obtains a sequence

```
Subset American human <[0.42278, 0.550354], 0.9, 10>
Subset American human <[0.435878, 0.563317], 0.9, 10>
Subset American human <[0.429098, 0.555749], 0.9, 10>
..
```

This is circular inference: one is repeatedly drawing a conclusion from some premises, and then using the conclusion to revise the premises. The situation gets even worse if one includes term probability inference in the picture. Iterated inference involves multiple evidence-counting, and so the resulting truth values become meaningless. Note, however, that due to the nature of indefinite probabilities, the generation of meaningless truth values appears to be bounded, so after a few iterations no additional meaningless values are generated.

One way to deal with this problem is to introduce a mechanism called "inference trails," borrowed from NARS. This is not the only approach, and in a later chapter we will discuss an alternative strategy which we call "trail free inference." But in the present section we will restrict ourselves to the trail-based approach, which is what is utilized in our current primary PLN software implementation. In the trail-based approach, one attaches to an Atom a nonquantitative object called the "inference trail" (or simply the "trail"), defined as a partial list of the terms and relationships that have been used to influence the Atom's truth-value. For the trail mechanism to work properly, each Atom whose truth-value is directly influenced by inference must maintain a trail object. Atoms newly formed in an AI system via external input or via noninferential concept creation have empty trails. Atoms formed via inference have nonempty trails.

The rule for trail updating is simple: In each inference step, the trail of conclusion is appended with the trails of the premises, as well as the premises themselves. The intersection of the trails of two atoms *A* and *B* is then a measure of the extent to which *A* and *B* are based on overlapping evidence. If this intersection is zero, then this suggests that *A* and *B* are independent. If the intersection is nonzero, then using *A* and *B* together as premises in an inference step involves "circular inference."

As well as avoiding circular inference, the trail mechanism has the function of preventing a false increase of "weight of evidence" based on revising (merging) atoms whose truth-values were derived based on overlapping evidence. It can also be used for a couple of auxiliary purposes, such as providing information when the system needs to explain how it arrived at a given conclusion, and testing and debugging of inference systems.

The use of inference trails means that the choice of which inference to do is quite a serious one: each inference trajectory that is followed potentially rules out a bunch of other trajectories, some of which may be better! This leads to subtle issues relating to inference control, some of which will be addressed in later chapters – but which will not be dealt with fully in these pages because they go too far beyond logic per se.

In practice, it is not feasible to maintain infinitely long trails, so a trail length is fixed (say, 10 or 20). This means that in seeking to avoid circular inference and related ills, a reasoning system like PLN can't use precise independence information, but only rough approximations. Pei Wang has formulated a nice aphorism in this regard: "When circular inference takes place and the circle is large enough, it's called coherence." In other words, some large-circle circularity is inevitable in any mind, and does introduce some error, but there is no way to avoid it without introducing impossible computational resource requirements or else imposing overly draconian restrictions on which inferences can occur. Hence, the presence of undocumented interdependencies between different truth-value assessments throughout an intelligent system's memory is unavoidable. The undocumented nature of these dependencies is a shortcoming that prevents total rationality from occurring, but qualitatively, what it does is to infuse the set of relationships and terms in the memory with a kind of "conceptual coherence" in which each one has been analyzed in terms of the cooperative interactions of a large number of others.

5.12 The Rule of Choice

When the revision rule is asked to merge two different truth-value estimates, sometimes its best course may be to refuse. This may occur for two reasons:

1. Conflicting evidence: the two estimates both have reasonably high weight of evidence, but have very different strengths.
2. Circular inference violations: the rule is being asked to merge a truth-value T_1 with another truth-value T_2 that was derived in part based on T_1.

The name "Rule of Choice" refers to a collection of heuristics used to handle these cases where revision is not the best option. There is not really one particular inference rule here, but rather a collection of heuristics each useful in different cases. In the case of conflicting evidence, there is one recourse not available in the circular inference case: creating a distributional truth-value. This is not always the best approach; its appropriateness depends on the likely reason for the existence of the conflicting evidence. If the reason is that the different sources observed different examples of the relationship, then a distributional truth-value is likely the right solution. If the reason is that the different sources simply have different opinions, but one and only one will ultimately be correct, then a distributional truth-value is likely not the right solution.

For instance, suppose we have two human Mars explorers reporting on whether or not Martians are midgets. If one explorer says that Martians are definitely midgets and another explorer says that Martians are definitely not midgets, then the relevant question becomes whether these explorers observed the same Martians or not. If they observed different Martians, then revision may be applied, but to preserve information it may also be useful to create a distributional truth-value indicating that midget-ness among Martians possesses a bimodal distribution (with some total midgets, the ones observed by the first explorer; and with some total non-midgets, the ones observed by the second explorer). On the other hand, if the two explorers saw the same Martians, then we may not want to just revise their estimates into a single number (say, a .5 strength of midget-ness for Martians); and if we do, we want to substantially decrease the weight of evidence associated with this revised strength value. Nor, in this case, do we want to create a bimodal truth-value: because the most likely case is that the Martians either are or are not midgets, but one of the explorers was mistaken.

On the other hand, one option that holds in both the conflicting evidence and circular inference cases is for the inference system to keep several versions of a given Atom around – distinct only in that their truth-values were created via different methods (they came from different sources, or they were inferentially produced according to overlapping and difficult-to-revise inference trails). This solution may be achieved knowledge-representation-wise by marking the different versions of an Atom as Hypothetical (see the following chapter for a treatment of Hypothetical relationships), using a special species of Hypothetical relationship that allows complete pass-through of truth-values into regular PLN inference.

Next we describe several approaches that are specifically applicable to the case where the Rule of Choice is used to avoid circular inference (i.e., where it is used to mediate trail conflicts). First, an accurate but terribly time-consuming approach is to take all the elements from the trails of the two relationships

```
Rel S P <t1>
Rel S P <t2>
```

merge them together into an atom set *S*, and then do a number of inference steps intercombining the elements of *S*, aimed at getting a new truth-value *t* so that

```
Rel S P <t>
```

This very expensive approach can only be taken when the relationship involved is very, very important. Alternately, suppose that

```
T = (Rel S P ).Trail <t1> MINUS (Rel S P).Trail <t2>
```

Then we can define

```
Rel S P <t3>
```

as the result of doing inference from the premises

```
Rel S P <t2>
```

```
T
```

and we can merge

```
Rel S P <t1>
Rel S P <t3>
```

using ordinary revision. This of course is also costly and only applicable in cases where the inference in question merits a fairly extreme allocation of computational resources.

Finally, a reasonable alternative in any Rule of Choice situation (be it conflicting evidence or circular inference related), in the case where limited resources are available, is to simply do revision as usual, but set the degrees of freedom of the conclusion equal to the maximum of the degrees of freedom of the premises, rather than the sum. This prevents the degrees of freedom from spuriously increasing as a result of false addition of overlapping evidence, but also avoids outright throwing-out of information. A reasonable approach is to use this crude alternative as a default, and revert to the other options on rare occasions when the importance of the inference involved merits such extensive attention allocation.

Chapter 6: First-Order Extensional Inference with Indefinite Truth-Values

Abstract In this chapter we exemplify the use of indefinite probabilities in a variety of PLN inference rules (including exact and heuristic ones).

6.1 Inference Using Indefinite Probabilities

In this chapter we show how the inference rules described above can be applied when truth-values are quantified as indefinite probabilities, rather than as individual probabilities or (probability, weight of evidence) pairs. We outline a general three-step process according to which a PLN inference formula, defined initially in terms of individual probabilities, may be evaluated using indefinite probabilities. We will first present these three steps in an abstract way; and then, in the following section, exemplify them in the context of specific probabilistic inference formulas corresponding to inference rules like term logic and modus ponens deduction, Bayes' rule, and so forth.

In general, the process is the same for any inference formula that can be thought of as taking in premises labeled with probabilities, and outputting a conclusion labeled with a probability. The examples given here will pertain to single inference steps, but the same process may be applied holistically to a series of inference steps, or most generally an "inference tree/DAG" of inference steps, each building on conclusions of prior steps.

Step One in the indefinite probabilistic inference process is as follows. Given intervals $[L_i, U_i]$ of mean premise probabilities, we first find a distribution from the "second-order distribution family" supported on $[L1_i, U1_i] \supset [L_i, U_i]$, so that these means have $[L_i, U_i]$ as $(100 \cdot b_i)\%$ credible intervals. The intervals $[L1_i, U1_i]$ either are of the form $\left[\frac{m}{n+k}, \frac{m+k}{n+k}\right]$, when the "interval-type" parameter associated with the premise is asymmetric, or are such that both of the intervals $[L1_i, L_i]$ and $[U_i, U1_i]$ each have probability mass $b_i/2$, when interval-type is symmetric.

Next, in Step Two, we use Monte Carlo methods on the set of premise truth-values to generate final distribution(s) of probabilities as follows. For each premise we randomly select n_1 values from the ("second-order") distribution found in Step One. These n_1 values provide the values for the means for our first-order distributions. For each of our n_1 first-order distributions, we select n_2 values to represent the first-order distribution. We apply the applicable inference rule (or tree of inference rules) to each of the n_2 values for each first-order distribution to generate the final distribution of first-order distributions of probabilities. We calculate the mean of each distribution and then – in Step Three of the overall process – find a $(100\cdot b_i)\%$ credible interval, $[L_f, U_f]$, for this distribution of means.

When desired, we can easily translate our final interval of probabilities, $[L_f, U_f]$, into a final, (s_f, n_f, b_f) triple of strength, count, and credibility levels, as outlined above.

Getting back to the "Bayesianism versus frequentism" issues raised in Chapter 4, if one wished to avoid the heuristic assumption of decomposability regarding the first-order plausibilities involved in the premises, one would replace the first-order distributions assumed in Step Two with Dirac delta functions, meaning that no variation around the mean of each premise plausibility would be incorporated. This would also yield a generally acceptable approach, but would result in overall narrower conclusion probability intervals, and we believe would in most cases represent a step away from realism and robustness.

6.1.1 The Procedure in More Detail

We now run through the above three steps in more mathematical detail.

In Step One of the above procedure, we assume that the mean strength values follow some given initial probability distribution $g_i(s)$ with support on the interval $[L1_i, U1_i]$. If the interval-type is specified as asymmetric, we perform a search until we find values *k2* so that $\int_{L_i}^{U_i} g_i(s)ds = b$, where $L_i = \frac{m_i}{n_i + k2}$ and $U_i = \frac{m_i + k2}{n_i + k2}$. If the interval-type is symmetric, we first ensure, via parameters, that each first-order distribution is symmetric about its mean s_i, by setting $L_i = s_i - d$, $U_i = s_i + d$ and performing a search for d to ensure that $\int_{L_i}^{U_i} g_i(s)ds = b$. In either case, each of the intervals $[L_i, U_i]$ will be a $(100\cdot b)\%$ credible interval for the distribution $g_i(s)$.

We note that we may be able to obtain the appropriate credible intervals for the distributions $g_i(s)$ only for certain values of b. For this reason, we say that a value b is truth-value-consistent whenever it is feasible to find $(100\cdot b)\%$ credible intervals of the appropriate type.

In Step Two of the above procedure, we create a family of distributions, drawn from a pre-specified set of distributional forms, and with means in the intervals $[L_i, U_i]$. We next apply Monte Carlo search to form a set of randomly chosen "premise probability tuples". Each tuple is formed via selecting, for each premise of the inference rule, a series of points drawn at random from randomly chosen distributions in the family. For each randomly chosen premise probability tuple, the inference rule is executed. And then, in Step Three, to get a probability value s_f for the conclusion, we take the mean of this distribution. Also, we take a credible interval from this final distribution, using a pre-specified credibility level b_f, to obtain an interval for the conclusion $[L_f, U_f]$.

6.2 A Few Detailed Examples

In this section we report some example results obtained from applying the indefinite probabilities approach in the context of simple inference rules, using both symmetric and asymmetric interval-types.

Comparisons of results on various inference rules indicated considerably superior results in all cases when using the symmetric intervals. As a result, we will report results for five inference rules using the symmetric rules; while we will report the results using the asymmetric approach for only one example (term logic deduction).

6.2.1 A Detailed Beta-Binominal Example for Bayes' Rule

First we will treat Bayes' rule, a paradigm example of an uncertain inference rule -- which however is somewhat unrepresentative of inference rules utilized in PLN, due to its non-heuristic, exactly probabilistic nature.

The beta-binomial model is commonly used in Bayesian inference, partially due to the conjugacy of the beta and binomial distributions. In the context of Bayes' rule, Walley develops an imprecise beta-binomial model (IBB) as a special case of an imprecise Dirichlet model (IDM). We illustrate our indefinite probabilities approach as applied to Bayes' rule, under the same assumptions as these other approaches. We treat here the standard form for Bayes' rule:

$$P(A|B) = \frac{P(A)P(B|A)}{P(B)}.$$

Consider the following simple example problem. Suppose we have 100 gerbils of unknown color; 10 gerbils of known color, 5 of which are blue; and 100 rats of known color, 10 of which are blue. We wish to estimate the probability of a randomly chosen blue rodent being a gerbil.

The first step in our approach is to obtain initial probability intervals. We obtain the following sets of initial probabilities shown in Tables 1 and 2, corresponding to credibility levels b of 0.95 and 0.982593, respectively.

Table 1. Intervals for credibility level 0.90		
EVENT	$[L,U]$	[L1, U1]
G	$\left[\frac{10}{21},\frac{12}{21}\right]$=[.476190, 0.571429]	[0.419724, 0.627895]
R	$\left[\frac{8}{21},\frac{12}{21}\right]$=[0.380952, 0.571429]	[0.26802, 0.684361]
B|G	[0.3, 0.7]	[0.0628418, 0.937158]
B|R	[0.06, 0.14]	[0.0125683, 0.187432]

Table 2. Intervals for credibility level 0.95		
EVENT	**[L, U]**	**[L1, U1]**
G	$\left[\frac{10}{21},\frac{12}{21}\right]$=[.476190, 0.571429]	[0.434369, 0.61325]
R	$\left[\frac{8}{21},\frac{12}{21}\right]$=[0.380952, 0.571429]	[0.29731, 0.655071]
B|G	[0.3, 0.7]	[0.124352, 0.875649]
B|R	[0.06, 0.14]	[0.0248703, 0.17513]

We begin our Monte Carlo step by generating n_1 random strength values, chosen from Beta distributions proportional to $x^{ks-1}(1-x)^{k(1-s)-1}$ with mean values of $s=\frac{11}{21}$ for *P*(G); $s=\frac{10}{21}$ for *P*(R); $s=\frac{1}{2}$ for $P(B|G)$; and $s=\frac{1}{10}$ for $P(B|R)$, and with support on [L1, U1]. Each of these strength values then serves, in turn, as parameters of standard Beta distributions. We generate a random sample of n_2 points from each of these standard Beta distributions.

We next apply Bayes' theorem to each of the $n_1 n_2$ quadruples of points, generating n_1 sets of sampled distributions. Averaging across each distribution then gives a distribution of final mean strength values. Finally, we transform our final distribution of mean strength values back to (*s*, *n*, *b*) triples.

6.2.1.1 Experimental Results

Results of our Bayes' rule experiments are summarized in Tables 3 and 4.

Table 3. Final probability intervals and strength values for $P(G|B)$ using initial b-values of 0.90

CREDIBILITY LEVEL	INTERVAL
0.90	[0.715577, 0.911182]
0.95	[0.686269, 0.924651]

Table 4. Final probability intervals and strength values for P(G|B) using initial b-values of 0.95

CREDIBILITY LEVEL	INTERVAL
0.90	[0.754499, 0.896276]
0.95	[0.744368, 0.907436]

6.2.1.2 Comparisons to Standard Approaches

It is not hard to see, using the above simple test example as a guide, that our indefinite probabilities approach generalizes both classical Bayesian inference and Walley's IBB model. First note that single distributions can be modeled as envelopes of distributions with parameters chosen from uniform distributions. If we model *P*(B) as a uniform distribution; *P*(A) as a single beta distribution and $P(B|A)$ as a single binomial distribution, then our method reduces to usual Bayesian inference. If, on the other hand, we model *P*(B) as a uniform distribution; *P*(A) as an envelope of beta distributions; and $P(B|A)$ as an envelope of binomial distributions, then our envelope approach reduces to Walley's IBB model. Our envelope-based approach thus allows us to model *P*(B) by any given family of distributions, rather than restricting us to a uniform distribution. This allows for more flexibility in accounting for known, as well as unknown, quantities.

To get a quantitative comparison of our approach with these others, we modeled the above test example using standard Bayesian inference as well as Walley's IBB model. To carry out standard Bayesian analysis, we note that given that there are 100 gerbils whose blueness has not been observed, we are dealing with 2^{100}

"possible worlds" (i.e., possible assignments of blue/non-blue to each gerbil). Each of these possible worlds has 110 gerbils in it, at least 5 of which are blue, and at least 5 of which are non-blue.

For each possible world w, we can calculate the probability that drawing 10 gerbils from the population of 110 existing in world W yields an observation of 5 blue gerbils and 5 non-blue gerbils. This probability may be written $P(D|H)$, where D is the observed data (5 blue and 5 non-blue gerbils) and H is the hypothesis (the possible world W).

Applying Bayes' rule, we have $P(H|D) = \frac{P(D|H)P(H)}{P(D)}$. Assuming that $P(H)$ is constant across all possible worlds, we find that $P(H|D)$ is proportional to $P(D|H)$. Given this distribution for the possible values of the number of blue gerbils, one then obtains a distribution of possible values $P(gerbil|blue)$ and calculates a credible interval. The results of this Bayesian approach are summarized in Table 5

Table 5. Final Probability Intervals for P(G|B)

CREDIBILITY LEVEL	INTERVAL
0.90	[0.8245614035, 0.8630136986]
0.95	[0.8214285714, 0.8648648649]

We also applied Walley's IBB model to our example, obtaining (with k=10) the interval $\left[\frac{2323}{2886}, \frac{2453}{2886}\right]$, or approximately [0.43007, 0.88465]. In comparison, the hybrid method succeeds at maintaining narrower intervals, albeit at some loss of credibility.

With a k-value of 1, on the other hand, Walley's approach yields an interval of [0.727811, 0.804734]. This interval may seem surprising because it does not include the average given by Bayes' theorem. However, it is sensible given the logic of Walley's approach. In this approach, we assume no prior knowledge of $P(G)$ and we have 10 new data points in support of the proposition that $P(G|B)$ is 55/65. So we assume beta distribution priors with s=0 and s=1 for the "endpoints" of $P(G)$ and use n=10 and p=11/13 for the binomial distribution for $P(G|B)$.

The density function thus has the form

$$f(x) = \frac{x^{ks}(1-x)^{k(1-s)}x^{np}(1-x)^{n(1-p)}}{\int_0^1 x^{ks}(1-x)^{k(1-s)}x^{np}(1-x)^{n(1-p)}dx}$$

Now for s=1 and the value of the learning parameter k=1, the system with no prior knowledge starts with the interval [1/3, 2/3]. With only 10 data points in support of p=11/13 and prior assumption of no knowledge or prior p=1/2, Walley's method is (correctly according to its own logic) reluctant to move quickly in support of p=11/13, without making larger intervals via larger k-values.

6.2.2 A Detailed Beta-Binominal Example for Deduction

Next we consider another inference rule, term logic deduction, which is more interesting than Bayes' rule in that it combines probability theory with a heuristic independence assumption. The independence-assumption-based PLN deduction rule, as derived in Chapter 5, has the following form for "consistent" sets A, B, and C:

$$s_{AC} = s_{AB}s_{BC} + \frac{(1-s_{AB})(s_C - s_B s_{BC})}{1-s_B}$$

where

$$s_{AC} = P(C|A) = \frac{P(A \cap C)}{P(A)}$$

assuming the given data $s_{AB} = P(B|A)$, $s_{BC} = P(C|B)$, $s_A = P(A)$, $s_B = P(B)$, and $s_C = P(C)$.

Our example for the deduction rule will consist of the following premise truth-values. In Table 6, we provide the values for $[L,U]$, and b, as well as the values corresponding to the mean s and count n. In the following sets of examples we assumed k=10 throughout.

Table 6. Premise truth-values used for deduction with symmetric intervals

Premise	s	$[L,U]$	$[L1, U1]$
A	11/23	[10/23, 12/23]≈[0.434783, 0.521739]	[0.383226, 0.573295]
B	0.45	[0.44, 0.46]	[0.428142, 0.471858]
AB	0.413043	[0.313043, 0.513043]	[0.194464, 0.631623]
BC	8/15	[7/15, 9/15] ≈ [0.466666, 0.6]	[0.387614, 0.679053]

We now vary the premise truth-value for variable C, keeping the mean s_C constant, in order to study changes in the conclusion count as the premise width $[L,U]$ varies. In Table 7, $b = 0.9$ for both premise and conclusion, and $s_C = 0.59$. The final count n_{AC} is found via the inverse Bernoulli function approach of Chapter 4.

Table 7. Deduction rule results using symmetric intervals

Premise *C*		**Conclusion *AC***		
[L,U]	***[L1, U1]***	**sAC**	***[L,U]***	***nAC***
[0.44, 0.74]	[0.26213, 0.91787]	0.575514	[0.434879, 0.68669]	11.6881
[0.49, 0.69]	[0.37142, 0.80858]	0.571527	[0.47882, 0.65072]	18.8859
[0.54, 0.64]	[0.48071, 0.69929]	0.571989	[0.52381, 0.612715]	42.9506
[0.58, 0.60]	[0.56814, 0.61186]	0.571885	[0.4125, 0.6756]	10.9609

For comparison of using symmetric intervals versus asymmetric, we also tried identical premises for the means using the asymmetric interval approach. In so doing, the premise intervals $[L, U]$ and $[L1, U1]$ are different as shown in Tables 8 and 9, using $b = 0.9$ as before.

Table 8. Premise truth-values used for deduction with asymmetric intervals

Premise	***s***	***[L,U]***	***[L1, U1]***
A	11/23	[0.44, 0.52]	[0.403229, 0.560114]
B	0.45	[0.44, 0.462222]	[0.42818, 0.476669]
AB	0.413043	[0.38, 0.46]	[0.3412, 0.515136]
BC	8/15	[0.48, 0.58]	[0.416848, 0.635258]

Table 9. Deduction rule results using symmetric intervals

Premise *C*		**Conclusion *AC***		
[L,U]	***[L1, U1]***	**sAC**	***[L,U]***	***nAC***
[0.40, 0.722034]	[0.177061, 0.876957]	0.576418	[0.448325, 0.670547]	13.683
[0.45, 0.687288]	[0.28573, 0.801442]	0.577455	[0.461964, 0.661964]	15.630
[0.50, 0.652543]	[0.394397, 0.725928]	0.572711	[0.498333, 0.628203]	26.604
[0.55, 0.617796]	[0.526655, 0.600729]	0.568787	[0.4125, 0.6756]	10.961

6.2.3 Modus Ponens

Another important inference rule is modus ponens, which is the form of deduction standard in predicate logic rather than term logic. Term logic deduction as described above is preferable from an uncertain inference perspective, because it generally supports more certain conclusions. However, the indefinite probabilities approach can also handle modus ponens; it simply tends to assign conclusions fairly wide interval truth-values.

The general form of modus ponens is

A
A → B
|-
B

To derive an inference formula for this rule, we reason as follows. Given that we know *P(A)* and *P(B|A)*, we know nothing about *P(B|¬A)*. Hence *P(B)* lies in the interval

$$[Q,R] = [P(A \text{ and } B), 1 - (P(A) - P(A \text{ and } B))]$$
$$= [P(B|A)P(A), 1 - P(A) + P(B|A)P(A)].$$

For the modus ponens experiment reported here we used the following premises: For A we used $\langle [L,U],b\rangle = \left\langle \left[\frac{10}{23},\frac{12}{23}\right],0.9\right\rangle$; and for A → B we used <[*L*,*U*],*b*>=<[0.313043,0.513043],0.9>. We proceed as usual, choosing distributions of distributions for both *P(A)* and *P(B|A)*. Combining these we find a distribution of distributions [*Q*,*R*] as defined above. Once again, by calculating means, we end up with a distribution of [*Q*,*R*] intervals. Finally, we find an interval [*L*,*U*] that contains (100·*b*)% of the final [*Q*,*R*] intervals. In our example, our final [*L*,*U*] interval at the *b* = 0.9 level is [0.181154, 0.736029].

6.2.4 Conjunction

Next, the AND rule in PLN uses a very simple heuristic probabilistic logic formula: $P(A \text{ AND } B) = P(A)P(B)$. To exemplify this, we describe an experiment consisting of assuming a truth-value of <[*L*, *U*], *b*>=<[0.4, 0.5], 0.9> for *A* and a truth-value of <[*L*, *U*], *b*>=<[0.2, 0.3], 0.9> for *B*.

The conclusion truth-value for *P(A* AND *B)* then becomes <[*L*,*U*],*b*>=<[0,794882, 0.123946], 0.9>.

6.2.5 Revision

Very sophisticated approaches to belief revision are possible within the indefinite probabilities approach; for instance, we are currently exploring the possibility of integrating entropy optimization heuristics as described in (Kern-Isberner, , 2004) into PLN for this purpose (Goertzel, 2008). At present, however, we treat revision with indefinite truth-values in a manner analogous to the way we treat them with simple truth-values. This approach seems to be effective in practice, although it lacks the theoretical motivation of the entropy minimization approach.

Suppose D1 is the second-order distribution for premise 1 and D2 is the second-order distribution for D2. Suppose further that n_1 is the count for premise 1 and n_2 is the count for premise 2. Let $w_1 = n_1/(n_1+n_2)$ and $w_2 = n_2/(n_1+n_2)$, and then form the conclusion distribution D = w_1 D1 + w_2D2. We then generate our <[*L*,*U*],*b*,*k*> truth-value as usual.

As an example, consider the revision of the following two truth-values <[0.1, 0.2], 0.9, 20> and <[0.3, 0.7], 0.9, 20>. Calculating the counts using the inverse function discussed in Chapter 4 gives count values of 56.6558 and 6.48493 respectively. Fusing the two truth-values yields <[0.13614, 0.233208], 0.9, 20> with a resulting count value of 52.6705.

The inclusion of indefinite probabilities into the PLN framework allows for the creation of a logical inference system that provides more general results than Bayesian inference, while avoiding the quick degeneration to worst-case bounds inherent with imprecise probabilities. On more heuristic PLN inference rules, the indefinite probability approach gives plausible results for all cases attempted, as exemplified by the handful of examples presented in detail here.

What we have done in this chapter is present evidence that the indefinite probabilities approach may be useful for artificial general intelligence systems – first because it rests on a sound conceptual and semantic foundation; and second because when applied in the context of a variety of PLN inference rules (representing modes of inference hypothesized to be central to AGI), it consistently gives intuitively plausible results, rather than giving results that intuitively seem too indefinite (like the intervals obtained from Walley's approach, which too rapidly approach [0,1] after inference), or giving results that fail to account fully for premise uncertainty (which is the main issue with the standard Bayesian or frequentist first-order-probability approach).

Chapter 7: First-Order Extensional Inference with Distributional Truth-Values

Abstract In this chapter we extend some of the ideas of the previous chapters to deal with a different kind of truth-value, the "distributional truth-value," in which the single strength value is replaced with a whole probability distribution. We show is that if discretized, step-function distributional truth-values are used, then PLN deduction reduces simply to matrix multiplication, and PLN inversion reduces to matrix inversion.

7.1 Introduction

The strength formulas discussed above have concerned SimpleTruthValues and IndefiniteTruthValues. However, the PLN framework as a whole is flexible regarding truth-values and supports a variety of different truth-value objects. In this chapter we extend some of the ideas of the previous sections to deal with a different kind of truth-value, the "distributional truth-value," in which the single strength value is replaced with a whole probability distribution. In brief, what we show is that if discretized, step-function distributional truth-values are used, then PLN deduction reduces simply to matrix multiplication, and PLN inversion reduces to matrix inversion.

Other extensions beyond this are of course possible. For instance, one could extend the ideas in this chapter to a SecondOrderDistributionalTruthValue, consisting of a completely-specified second-order probability distribution (as opposed to an indefinite probability, which is a sketchily specified, compacted second-order probability distribution). But these and additional extensions will be left for later work.

Before launching into mathematics, we will review the conceptual basis for "distributional truth-values" by elaborating an example. Given a "man" Atom and a "human" Atom, suppose we've observed 100 humans and 50 of them are men. Then, according to the SimpleTruthValue approach, we may say

```
Subset man human < 1>
Subset human man <.5>
```

What this says is: All observed examples of "man" are also examples of "human"; half of the examples of "human" are also examples of "man."

The limitations of this simple approach to truth-value become clear when one contrasts the latter of the two relationships with the relation

```
Subset human ugly <.5>
```

Suppose that from listening to people talk about various other people, an AI system has gathered data indicating that roughly 50% of humans are ugly (in terms of being called ugly by a majority of humans talking about them). The difference between this observation and the observation that 50% of humans are men is obvious:

- In the case of man-ness, nearly 50% of humans totally have it, and nearly 50% of humans totally don't have it; intermediate cases are very rare
- In the case of ugliness, it's not the case that nearly 50% of humans are not at all ugly whereas nearly 50% are totally ugly. Rather there is a spectrum of cases, ranging from no ugliness to absolute ugliness, with the majority of people close to the middle (just barely uglier than average, or just barely less ugly than average).

In the language of statistics, man-ness has a *bimodal* distribution, whereas ugliness has a *unimodal* distribution. Taking this difference in underlying distribution into account, one can do more refined inferences.

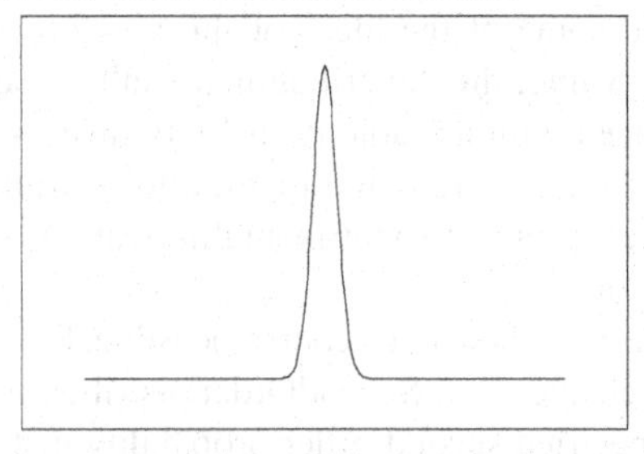

Example of Unimodal Distribution

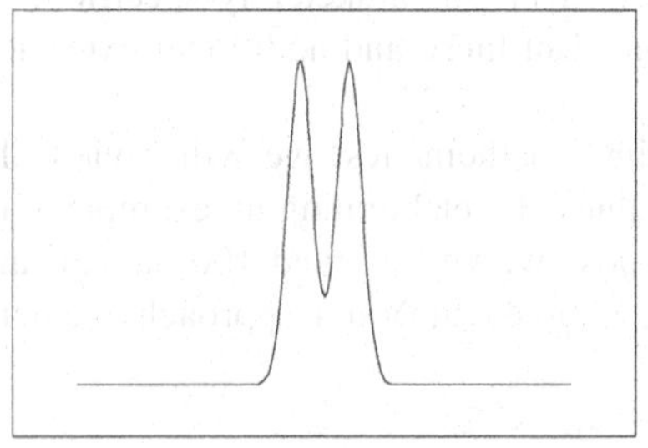

Example of Bimodal Distribution

From examples like this, one arrives at the idea of "distributional truth-values" – TruthValue objects that contain not only a strength value (reflecting the mean of a truth-value distribution) but a better approximation to an entire probability distribution reflecting the truth-value of an entity. We have experimented with both StepFunctionTruthValue and PolynomialTruthValue objects, two specific implementations of the notion of distributional truth-values.

The "distributions" involved are functions $f : [0,1]^2 \rightarrow [0,1]$, where for example, if

```
Subset A C <f,w>
```

then

$$f(x,y) = t$$

means

$$P(\text{ Member Q C } < y > | \text{ Member Q C } < x >) = t,$$

and w represents the weight-of-evidence.

In practice, one cannot look at the full space of distributions f, but must choose some finite-dimensional subspace; and as noted above, we have explored two possibilities: polynomials and step functions. Polynomials are more elegant and in principle may lead to more efficient computational algorithms, but the inference formulas are easier to work out for step functions, and the implementation complexity is also less for them, so after some preliminary work with polynomials we decided to focus on step function distributional approximations.

7.2 Deduction and Inversion Using Distributional Truth-Values

Now we give some mathematics regarding PLN inference with distributional truth-values, assuming discretized, step-function distributional truth-values. The mathematical framework used here is as follows. Consider three Terms, A, B and C, with Term probability distributions denoted by vectors p^A, p^B, p^C where, e.g.,

$$p^A = (p_1^A, p_2^A, \ldots, p_n^A)$$
$$p_i^A = P(A_i)$$

The event A_i is defined as

$$t_{i,A} < P(A) \le t_{i+1,A}$$

where $(t_{1,A} = 0, t_{2,A}, \ldots, t_{n,A=1})$ is a (not necessarily equispaced) partition of the unit interval.

It may often be useful to set the partition points so that

$$P_i^A \cong \frac{1}{n-1}$$

for all i. Note that, if one follows this methodology, the partition points will generally be unequal for the different Terms in the system. This is not problematic for any of the inference mathematics we will use here. In fact, as we'll see, deduction works perfectly well even if the number of partition points is different for different Terms. However, inversion becomes problematic if different Terms have different numbers of partition points; so it's best to assume that n is a constant across all Terms.

Next, define the conditional probability $P(C|A)$ by the matrix P^{AC}, where

$$P_{i,j}^{AC} = P(C_j|A_i)$$

So long as A and C have the same number of partition points, this will be a square matrix.

Given this set-up, the deduction rule is obtained as follows.

Theorem. *Assume* A_i *and* C_j *are independent in* B_k *for all i, j, k. Then* $P^{AC} = P^{AB}P^{BC}$.

Proof.

The theorem follows from the key equation

$$P(C_j|A_i) = \sum_{k=1}^{n} P(C_j|B_k)P(B_k|A_i)$$

which we formally demonstrate in a moment. This equation, after a change of notation, is

$$P_{i,j}^{AC} = \sum_{k=1}^{n} P_{i,k}^{AB} P_{k,j}^{BC}$$

which according to the definition of matrix multiplication means

$$P^{AC} = P^{AB} P^{BC}$$

Now, the key equation is demonstrated as follows. We know

$$P(C_j|A_i) = \sum_{k=1}^{n} P((C_j \cap B_k)|A_i) = \frac{\sum_{k=1}^{n} P(C_j \cap B_k \cap A_i)}{P(A_i)}$$

Assuming C_j and A_i are independent in B_k, we have

$$P(C_j|A_i) = \sum_{k=1}^{n} P(C_j|B_k) \frac{P(A_i \cap B_k)}{P(A_i)} = \sum_{k=1}^{n} P(C_j|B_k) P(B_k|A_i)$$

The critical step in here is the independence assumption

$$P(C_j \cap B_k \cap A_i) = P(C_j|B_k) P(A_i \cap B_k)$$

To see why this is true, observe that the assumption "A_i and C_j are independent in B_k" implies

$$|C_j \cap B_k \cap A_i| = P(C_j|B_k)|A_i \cap B_k|.$$

Dividing both sides by N, the size of the implicit universal set, yields the desired equation, completing the proof. **QED**

Next, the inversion rule is obtained from the equation

$$P^{B} = P^{AB} P^{A}$$

(which is obvious, without any independence assumptions), which implies

$$P^{A} = (P^{AB})^{-1} P^{B}$$

Thus, if a relationship (SubsetRelationship A B) has distributional TruthValue P^{AB}, it follows that the relationship (SubsetRelationship B A) should have distributional TruthValue $(P^{AB})^{-1}$. In other words, PLN inversion is matrix inversion.

Note that no term probabilities are used in any of these formulas. However, term probabilities can be derived from the distributional TruthValues. Specifically, the term probability for A is the expected value of the distributional truth-value for A.

Also, note that the inversion formula has error magnification problems due to the fact that the matrix inversion formula involves division by a determinant. And, in the unlikely case where the determinant of P^{AB} is zero, the inversion formula fails, though one may use matrix pseudoinverses (Weisstein, 2008) and probably still obtain meaningful results.

Continuous versions of the equations may also be derived, by shrinking the partition in the above formulas.

7.3 Relationship between SimpleTruthValue and DistributionalTruthValue Inference Formulas

In the case where n=2, one can obtain something quite similar (but not identical) to the SimpleTruthValue formulas by setting the partition interval $t_{1,A}$ for the Term A so that

$$\text{mean}(P(A)) = P(t_{1,A} < P(A) \leq t_{2,A} = 1)$$

and doing something similar for the Terms B and C. In this case one has, e.g.,

$$P(A) = (P(A), 1 - P(A)).$$

Regardless of how one sets the t_1 values, one can obtain fairly simple formulas for the 2-partition-interval case. One has, e.g.,

$$P^{AB} = \begin{bmatrix} P(B_1|A_1) & P(B_1|A_2) \\ P(B_2|A_1) & P(B_2|A_2) \end{bmatrix}$$

or equivalently

$$P^{AB} = \begin{bmatrix} P(B_1|A_1) & P(B_1|A_2) \\ 1 - P(B_1|A_1) & 1 - P(B_1|A_2) \end{bmatrix}$$

And for the inversion formula, one has

$$\begin{aligned} Det(P^{AB}) &= P(B_1|A_1) - P(B_1|A_1)P(B_1|A_2) - P(B_1|A_2) + P(B_1|A_2)P(B_1|A_1) \\ &= P(B_1|A_1) - P(B_1|A_2) \end{aligned}$$

so that inversion is only degenerate when $P\left(t_{1,B} < P(B)\right)$ is independent of whether "$t_{1,A} < P(A)$" is true or not. The inversion formula is then given by

$$\left[P(B_1|A_1) - P(B_1|A_2)\right]\left(P^{AB}\right)^{-1} = \begin{bmatrix} 1 - P(B_1|A_2) & -P(B_1|A_2) \\ -1 + P(B_1|A_1) & P(B_1|A_1) \end{bmatrix}$$

Note that here, as in general,

$$\left(P^{AB}\right)^{-1} \neq P^{BA}$$

even though both

$$P^A = \left(P^{AB}\right)^{-1} P^B$$
$$P^A = P^{BA} P^B$$

hold. This is not surprising, as many different 2D matrices map a particular 1D subspace of vectors into another particular 1D subspace (the subspace here being the set of all 2D vectors whose components sum to 1).

The term most analogous to ordinary PLN-FOI inversion is

$$P(B_1) = \left(P^{AB}\right)^{-1}{}_{11} P(A_1) + \left(P^{AB}\right)^{-1}{}_{12} P(A_2)$$
$$= \frac{\left(1 - P(B_1|A_2)\right)P(A_1) - P(B_1|A_2)P(A_1)}{P(B_1|A_1) - P(B_1|A_2)}$$

Note that we are inverting the overall P^{AB} matrix without inverting the individual entries; i.e., because $\left(P^{AB}\right)^{-1} \neq P^{BA}$ we do not get from this an estimate of $P(A_1|B_1)$.

The term of the deduction formula that is most analogous to PLN-FOI deduction is given by

$$P_{11}^{AC} = P\left(P(C) > t_{1,C} \middle| P(A) > t_{1,A}\right)$$
$$= P(C_1|B_1)P(B_1|A_1) + P(C_1|B_2)P(B_2|A_1)$$

which is similar, but not equal, to the FOI formula

$$s_{AC} = P(C|B)P(B|A) + P(C|\neg B)P(\neg B|A).$$

Chapter 8: Error Magnification in Inference Formulas

Abstract In this chapter, we mathematically explore the sensitivity of PLN strength formulas to errors in their inputs.

8.1: Introduction

One interesting question to explore, regarding the PLN strength formulas, is the sensitivity of the formulas to errors in their inputs. After all, in reality the "true" probabilities s_A, s_B, s_C, s_{AB}, s_{BC} are never going to be known exactly. In practical applications, one is nearly always dealing with estimates, not exactly known values. An inference formula that is wonderful when given exact information, but reacts severely to small errors in its inputs, is not going to be very useful in such applications.

This issue is dealt with to a large extent via the weight of evidence formulas presented in Chapters 4 and 5. In cases where an inference formula is applied with premises for which it has high error sensitivity, the weight of evidence of the conclusion will generally suffer as a consequence. However, it is interesting from a theoretic standpoint to understand when this is going to occur – when error sensitivity is going to cause inferences to result in low-evidence conclusions. Toward that end, in this chapter we use the standard tools of multivariable calculus to study the sensitivity of the inference formulas with respect to input errors.

The main conceptual result of these calculations is that without careful inference control, a series of probabilistic inferences may easily lead to chaotic trajectories, in which the strengths of conclusions reflect magnifications of random errors rather than meaningful information. On the other hand, careful inference control can avoid this chaotic regime and keep the system in a productive mode.

The first, obvious observation to make regarding inferential error sensitivity in PLN is that if the errors in the inputs to the formulas are small enough the error in the output will also be small. That is, the inference formulas are continuous based on their premise truth values. In the case of deduction, for example, this conclusion is summarized by the following theorem:

Theorem (PLN Deduction Formula Sensitivity)

Let U denote a set with |U| elements. Let Sub(m) denote the set of subsets of U containing m elements. Let $\approx$ indicate an approximate inequality. Assume $s_B \neq 1$, and $s'_A \approx s_A$, $s'_B \approx s_B$, $s'_C \approx s_C$, $s'_{AB} \approx s_{AB}$, $s'_{BC} \approx s_{BC}$, and let

$$f(x) = P\left(P(C|A) = x \,\middle|\, \begin{bmatrix} A \in \text{Sub}(|U|s'_A), B \in \text{Sub}(|U|s'_B), C \in \text{Sub}(|U|s'_C), \\ P(B|A) = s'_{AB}, P(C|B) = s'_{BC} \end{bmatrix}\right).$$

Then, where E() denotes the expected value (mean), we have

$$E[f(x)] \approx s_{AC} = s_{AB}s_{BC} + (1 - s_{AB})(s_C - s_B s_{BC})/(1 - s_B)$$

More formally: $\forall \in > 0, \exists \delta > 0$ so that $|P(A) - s_A| < \in$, $|P(B) - s_B| < \in$, $|P(C) - s_C| < \in$, $|P(B|A) - s_{AB}| < \in$, and $|P(C|B) - s_{BC}| < \in$ implies $|E[f(x) - s_{AC}]| < \delta$ where s_{AC} is defined by the above formula. The proof of this theorem is obvious and is omitted.

The situation with abduction is fairly similar to that with deduction, except that there is an additional restriction: s_B cannot equal 0 or 1. This is achievable by appropriate adjustment of the count N of the universe.

On the other hand, a glance back at the induction formula will show that the situation is a little different there. For induction, to avoid sensitive dependence on input values we need s_B to be bounded away from 1, and we also need s_A to be bounded away from 0. If s_B is near 1 or s_A is near 0, induction should not be carried out, because the results can't be trusted. And, unlike in the deduction case, this problem can't always be avoided by a careful choice of universe size. The problem is that if one increases the universe size to make s_B smaller, one also makes s_A smaller, which may bring it too close to zero. If the term B is a lot more frequent than the term A, then there is no way to do inductive inference meaningfully, unless one is sure that all one's premise truth-values are known *very close to exactly*.

One interesting consequence of this fact is that, in the case of induction, there is no way to set the universe size globally that will make all inductions in the system even minimally reliable. To get reliable inductions, one *has* to let the universe size vary across different inferences. So, if one wants to use the inductive inference rule (a very valuable heuristic), one has to give up the idea of a single global "space of events" serving as an implicit universe for all inferences in the system. One has to consider inferences as contextual.

These observations about the singularities of the inference formulas, however, are only the most simplistic way of studying the formulas' error-sensitivity. A deeper view is obtained by looking at the partial derivatives of the inference formulas, which are shown below.

These calculations – to be discussed in more detail below – reveal that the degree of error-sensitivity is highly dependent on the particular input values. For instance, consider the deduction formula. For the input values $s_{BC} = .5$, $s_C = s_B = .25$, we have

$$\frac{\partial s_{AC}}{\partial s_{AB}} = \frac{1}{3}$$

which means that the deduction formula decreases rather than increases errors in its input s_{AB} value. On the other hand, if $s_{BC} = 1$, s_C=.25, s_B = .5, then we have

$$\frac{\partial s_{AC}}{\partial s_{AB}} = 1.5$$

which means that errors in the input s_{AB} value are increased by a factor of 1.5.

To really understand error magnification in PLN, one has to look at the norm of the gradient vectors of the inference rules, as will be discussed in the following section. These analyses basically confirm what the above partial-derivative examples suggest: depending on the input values, sometimes an inference rule will magnify error, and sometimes it will decrease error (and sometimes it'll leave the error constant).

If errors are magnified in the course of an inference, then we're essentially losing precision in the course of the inference – we're losing information. Let's say we have a set of strength values that are known to 10 bits of precision apiece. If we do a series of inference steps, each one of which magnifies error by a factor of 2, then after 10 steps of inference we'll be dealing with totally meaningless conclusions. We will have "chaos" in the series of strengths associated with the trail of inferences, in the sense that the strength values observed will draw on later and later bits of the initial probabilities. In a practical computational context there are only a few significant bits and the rest is roundoff error, which means that in some cases, after enough inference steps have passed, the strength values could come to consist of nothing but roundoff error. Fortunately, this potentially very serious problem can be managed via careful attention to weight of evidence.

In this chapter we will focus on the SimpleTruthValue case, but the issue of error magnification and inferential chaos arises equally severely in other cases. In the DistributionalTruthValue case, error magnification will often result in probability distribution function (PDF) truth-values that are relatively uniformly distributed across the interval.

8.2: Gradient Norms and Error Magnification

Our assessment of the error magnification properties of PLN rules relies on computations of the norms of the gradient vectors of the rules. The concepts of error magnification and chaos in a discrete dynamical systems context are covered well in Devaney's book (Devaney, 1989.)

Basic Mathematics of Gradient Norms

All the PLN strength formulas are of the form

$$v = f(w)$$

where

- w lies in n-dimensional space (4-dimensional for deduction, 3-dimensional for inversion)
- v lies in 1-dimensional space

Let's ask what happens if w is perturbed by a small error e, so that we're looking at

$$v' = f(w') = f(w + e)$$

We can write

$$v' = f(w) + ((grad(f))(w)) \bullet e + O(e^2)$$

where $grad(f) = f_1 + f_2 + \cdots + f_n$, and $f_i = \frac{\partial f}{\partial x_i}$.

Since we have

$$v' - v \approx ((grad(f)(w))) \bullet e$$

it follows that

$$\|v' - v\| \approx \|(grad(f)(w))\| \|w - w'\| \cos\theta|$$

and on average

$$\|v' - v\| \approx \frac{2}{\pi} \|(grad(f)(w))\| \|w - w'\|.$$

As a result, if we want the amount of error to decrease from one step to the next, we'd like to have

$$\|(grad(f)(w))\| < 1$$

and if we want the amount of error to decrease on average we definitely need

$$\left\|\left(grad(f)(w)\right)\right\| < \frac{\pi}{2},$$

(though having only this is chancy, because one could always find that the upper bound of the inequality was realized for a long enough time to cause real trouble).

8.2.2 Inference Formula Gradient Norms

The following section presents algebraic calculations and graphical depictions of the gradient norms for the deduction, inversion, sim2inh and inh2sim formulas. As the illustrations given here show, the gradient norms behave rather differently for the different rules.

For deduction the norm of the gradient is often, but by no means always, less than 1. For some parameter values it's usually greater than 1; for others it's nearly always less. The algebraic formula involved is complex, but one clear fact is that when s_B is close to 1, "good" gradient norms are relatively rare. A big universe size seems to push the rule toward stability.

We need to be more careful with the inversion rule. As s_A gets close to zero, the gradient norms increase toward infinity, due to division by s_A. As s_A approaches 1 another problem occurs in that the region of consistent inputs becomes very small. For values of s_A that are between these two extremes, on the other hand, sizable regions for s_B and s_{AB} exist for which the gradient norm is "good."

For inh2sim, the gradient norm maxes out around 1.4. The problem area occurs when both premises, s_{AC} and s_{CA}, are very close to 1; that is, when *A* and *C* almost entirely overlap. It would be a shame to avoid doing inh2sim in these cases, which suggests that a more subtle control strategy than "avoid all error magnification" is called for (a topic we'll return to below).

For sim2inh, the gradient norm can get quite large (20 or so). This tends to occur when s_B is large, and s_C and sim_{BC} are both small. If we take s_B very small, then the gradient norm tends to be bounded above by 1. Here, a large universe size makes this rule dampen rather than magnify error.

8.2.3 Visual Depiction of PLN Formula Derivatives

Now we report some Maple work deriving and displaying the gradient norms of the PLN inference formulas. The partial derivatives of the PLN deduction formula are computed as:

diff(dedAC(sA,sB,sC,sAB,sBC),sA);

$$0$$

diff(dedAC(sA,sB,sC,sAB,sBC),sB);

$$-\frac{(1-sAB)sBC}{1-sB}+\frac{(1-sAB)(sC-sBsBC)}{(1-sB)^2}$$

diff(dedAC(sA,sB,sC,sAB,sBC),sC);

$$\frac{1-sAB}{1-sB}$$

diff(dedAC(sA,sB,sC,sAB,sBC),sAB);

$$sBC-\frac{sC-sBsBC}{1-sB}$$

diff(dedAC(sA,sB,sC,sAB,sBC),sBC);

$$sAB-\frac{(1-sAB)sB}{1-sB}$$

8.2.3.1 Depictions of the Gradient Norm of the Deduction Formula

The norm of the gradient vector of the PLN deduction formula is thus

dedAC_grad_norm := (sA,sB,sC,sAB,sBC) -> ((-(1-sAB)*sBC/(1-sB)+(1-sAB)*(sC-sB*sBC)/(1-sB)^2)^2 + ((1-sAB)/(1-sB))^2 + (sBC-(sC-sB*sBC)/(1-sB))^2 + (sAB-(1-sAB)*sB/(1-sB))^ 2) ^ (1/2) *(Heaviside(sAB-max(((sA+sB-1)/sA),0))-Heaviside(sAB-min(1,(sB/sA))))*(Heaviside(sBC-max(((sB+sC-1)/sB),0))-Heaviside(sBC-min(1,(sC/sB))));

$$dedAC\_grad\_norm := (sA, sB, sC, sAB, sBC) \to \left(\left(-\frac{(1-sAB)(sC-sBsBC)}{(1-sB)^2} \right) + \frac{(1-sAB)^2}{(1-sB)^2} + \left(sBC - \frac{sC-sBsBC}{1-sB} \right)^2 + \left(sAB - \frac{(1-sAB)sB}{1-sB} \right)^2 \right)^{1/2}$$

Finally, we are ready to show graphs of the deduction formula. We display two sets of graphs:

plot3d(dedAC_grad_norm(.1,.1,.1, sAB, sBC), sAB=0..1, sBC=0..1, labels=[sAB,sBC,ded], numpoints=800, axes=BOXED);

is the graph of the norm of the gradient vector, while

dedAC_grad_norm_only01 := (sA, sB, sC, sAB, sBC) -> piecewise (dedAC_grad_norm (sA, sB, sC, sAB, sBC) < 0, 0, dedAC_grad_norm(sA, sB, sC, sAB,sBC) < 1, dedAC_grad_norm(sA, sB, sC, sAB, sBC), 0);

is a filtered version of the graph, showing only those values of the norm that are less than or equal to 1.

plot3d(dedAC_grad_norm(.1,.1,.1, sAB, sBC), sAB=0..1, sBC=0..1, labels=[sAB,sBC,ded], numpoints=800, resolution = 400, axes=BOXED);

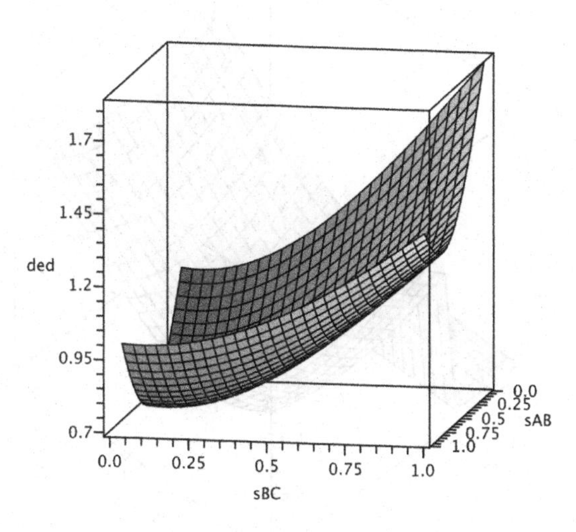

```
plot3d(dedAC_grad_norm_only01(.1,.1,.1,sAB,sBC),sAB=0..1,sBC=0..1,la
bels=[sAB,sBC,ded],numpoints=800,axes=BOXED);
```

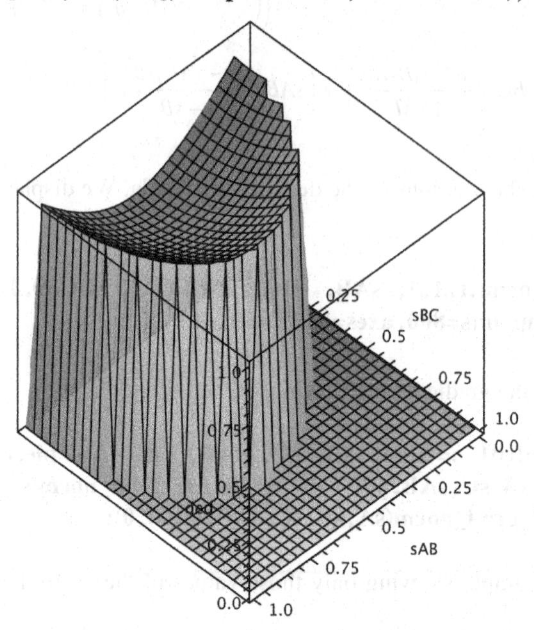


```
plot3d( dedAC_grad_norm(.1,.4, .7,sAB, sBC), sAB=0 ..1, sBC=0..1,
axes=BOXED);
```

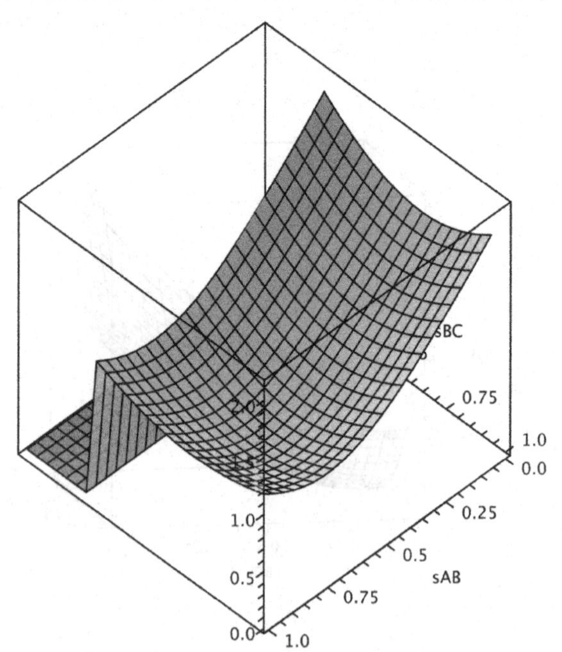

plot3d(dedAC_grad_norm_only01(.1,.4,.7,sAB,sBC),sAB=0..1,sBC=0..1,labels=[sAB,sBC,dedAC],numpoints=800,axes=BOXED);

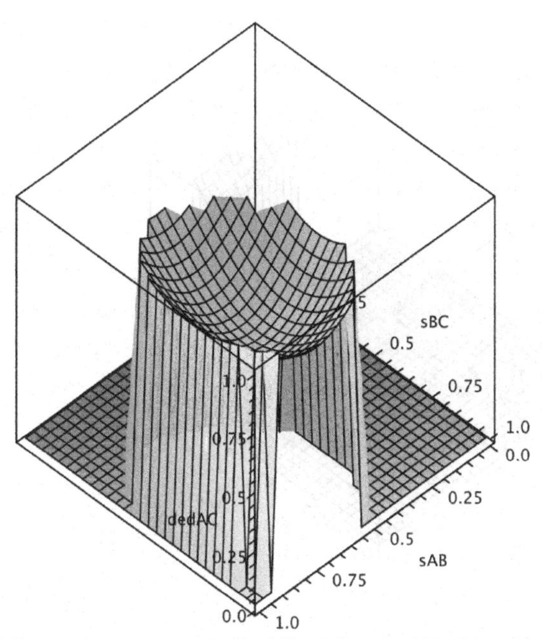

plot3d(dedAC_grad_norm(.4,.1,.9,sAB,sBC), sAB=0 ..1, sBC=0..1, axes=BOXED);

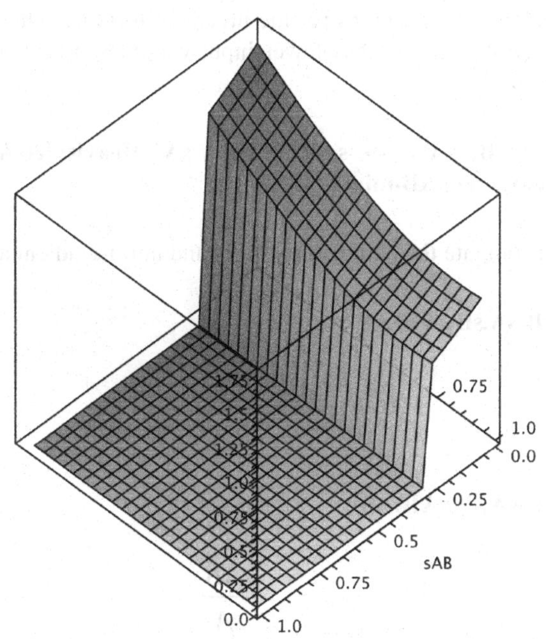

plot3d(dedAC_grad_norm_only01(.4,.1,.9,sAB,sBC),sAB=0..1,sBC=0..1,labels=[sAB,sBC,ded],numpoints=1000,axes=BOXED);

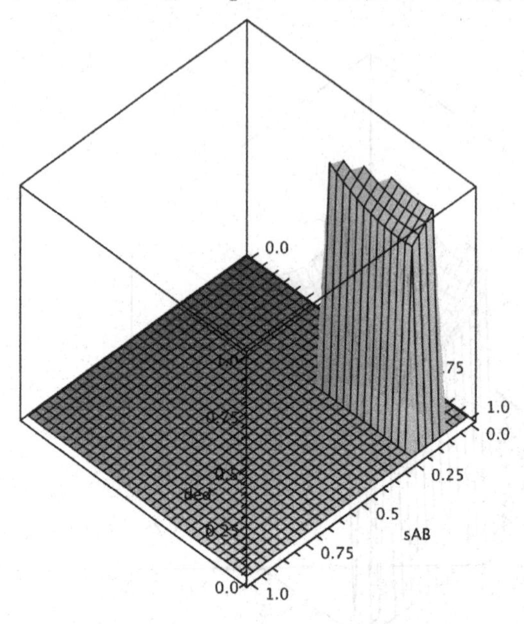

8.2.3.2 Depictions of the Gradient Norm of the Inversion Formula

Now we give similar figures for the inversion formula. These are simpler because the inversion formula has fewer input variables. Recall that the inversion formula looks like

invAB := (sA,sB, sAB) -> sAB * sB / sA*(Heaviside(sAB-max(((sA+sB-1)/sA),0))-Heaviside(sAB-min(1,(sB/sA))));

We may thus compute the partial derivatives and norm-gradient as

diff(invAB(sA,sB,sAB),sA);

$$-\frac{sABsB}{sA^2}$$

diff(invAB(sA,sB,sAB),sB);

$$\frac{sAB}{sA}$$

diff(invAB(sA,sB,sAB),sAB);

$$\frac{sB}{sA}$$

invAB_nd := (sA,sB, sAB) -> ((sAB/sA)^2 + (-sAB*sB/sA^2)^2 + (sB/sA) ^2)^(1/2)*(Heaviside(sAB-max(((sA+sB-1)/sA),0))-Heaviside(sAB-min(1,(sB/sA))));

$$invAB\_nd := (sA, sB, sAB) \rightarrow \sqrt{\frac{sAB^2}{sA^2} + \frac{sAB^2 sB^2}{sA^4} + \frac{sB^2}{sA^2}} \left(\mathrm{Heaviside}\left(sAB - \max\left(\frac{sA + sB - 1}{sA}, 0 \right) \right) - \mathrm{Heaviside}\left(\mathrm{sAB} - \min\left(1, \frac{\mathrm{sB}}{\mathrm{sA}} \right) \right) \right)$$

The following function returns the inversion derivative norm only in cases where it's in [0,1] (where we have error decrease rather than increase), and returns 0 otherwise:

invAB_nd_only01:=(sA,sB,sAB)->piecewise(invAB_nd(sA,sB,sAB)<0, 0,invAB_nd(sA,sB,sAB)<1,invAB_nd(sA,sB,sAB),0);

Now we give plots of inversion over all inputs, for various values of the input variable s_A.

plot3d(invAB_nd(.01,sB,sAB), sB=0 ..1, sAB=0..1, axes=BOXED, numpoints=1000, resolution = 400, labels=[sB,sAB,invAB_nd]);

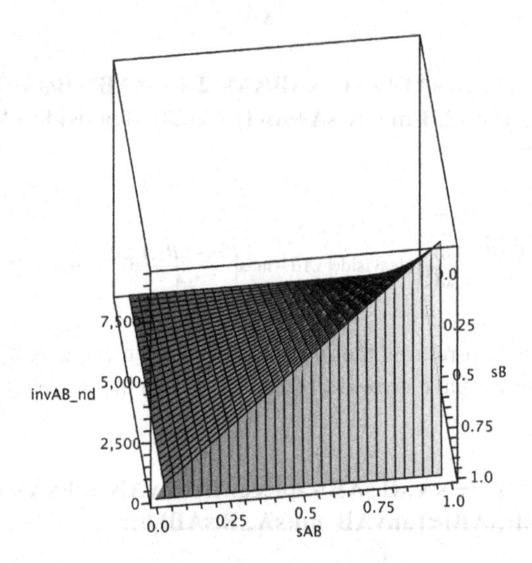

plot3d(invAB_nd(.1,sB,sAB), sB=0 ..1, sAB=0..1, axes=BOXED, numpoints=800, resolution=400, labels=[sB,sAB,invAB_nd]);

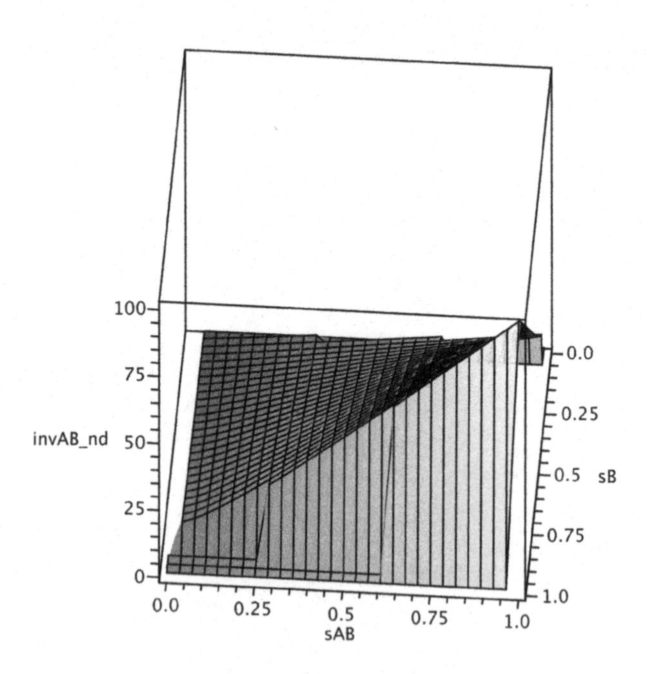

plot3d(invAB_nd_only01(.1,sB,sAB), sB=0.01 ..1, sAB=0..1, axes=BOXED, numpoints=1000, resolution=400, labels=[sB,sAB,invAB_nd]);

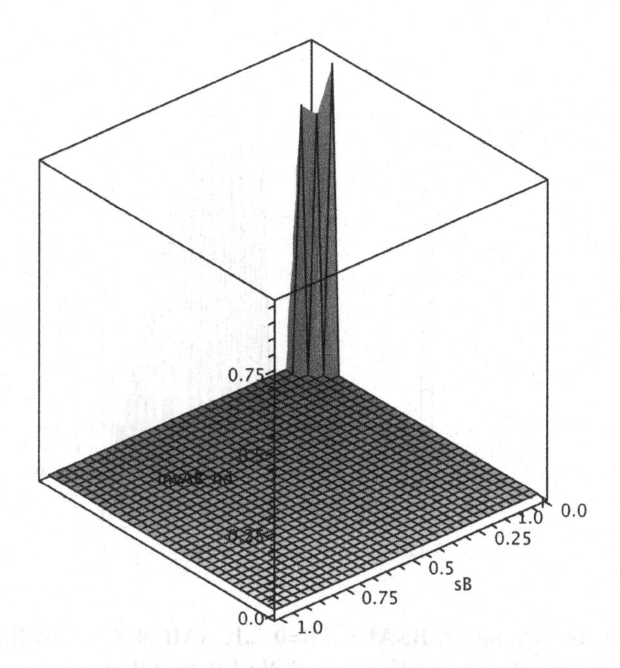

plot3d(invAB_nd(.5,sB,sAB), sB=0 ..1, sAB=0..1, axes=BOXED, numpoints=800, resolution = 400, labels=[sB,sAB,invAB_nd]);

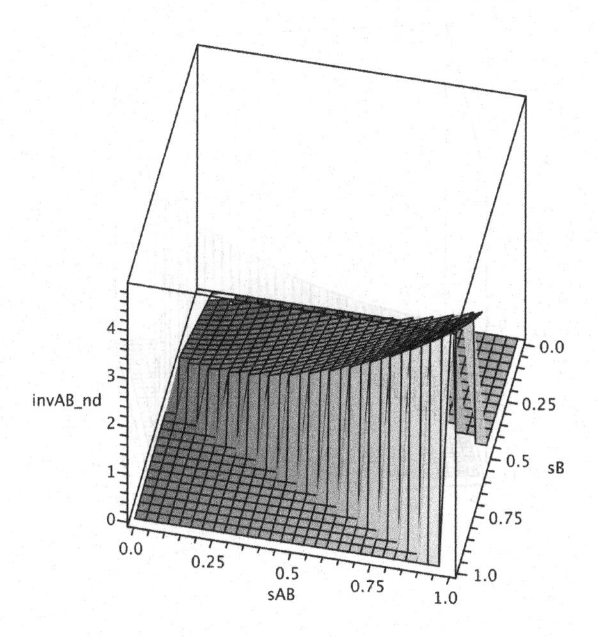

plot3d(invAB_nd_only01(.5,sB,sAB), sB=0 ..1, sAB=0..1, axes=BOXED, numpoints=1000, resolution = 400, labels=[sB,sAB,invAB_nd]);

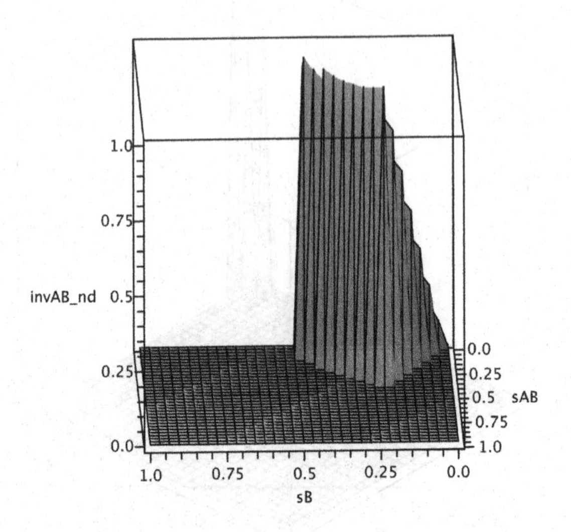

plot3d(invAB_nd(.9,sB,sAB), sB=0 ..1, sAB=0..1, axes=BOXED, numpoints=1000, resolution = 400, labels=[sB,sAB,invAB_nd]);

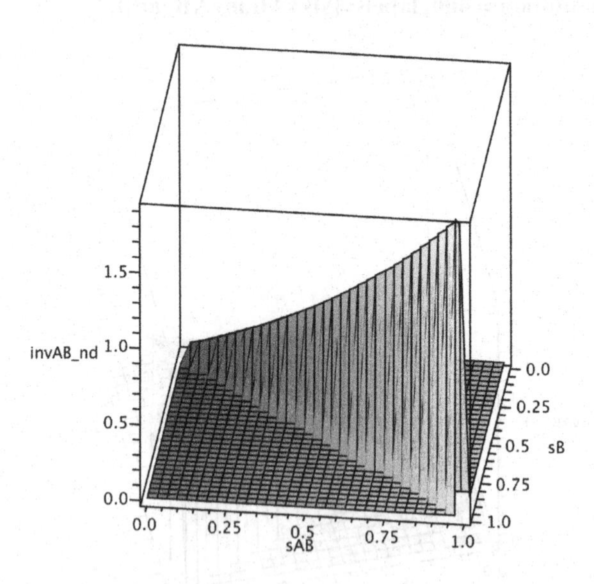

plot3d(invAB_nd_only01(.9,sB,sAB), sB=0 ..1, sAB=0..1, axes=BOXED, numpoints=1000, resolution = 400, labels=[sB,sAB,invAB_nd]);

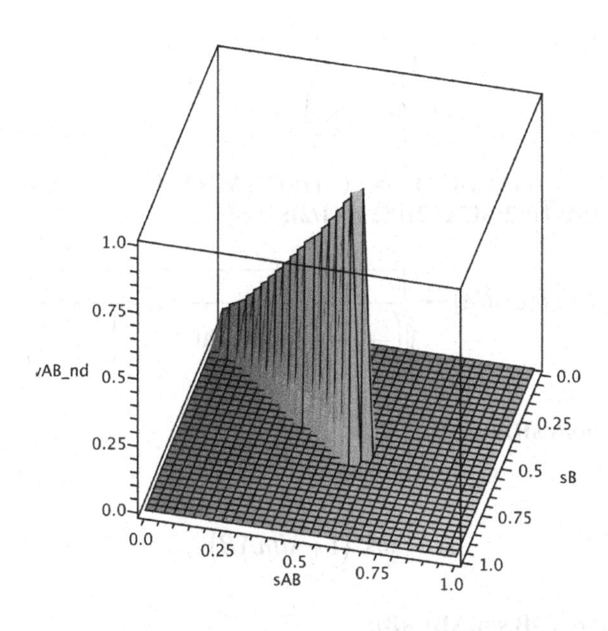

8.2.3.3 Depictions of the Gradient Norms of the Similarity-Inheritance Conversion Formulas

Finally, in this subsection we give a similar treatment for the sim2inh conversion formula. Here we have

sim2inh := (sA,sB,simAB) -> (1 + sB/sA) * simAB / (1 + simAB) *(Heaviside(simAB-max(((sA+sB-1)),0))-Heaviside(simAB-min(sA/sB, (sB/sA))));

inh2sim := (sAC,sCA) -> 1/(1/sAC + 1/sCA - 1);

from which we calculate the derivatives

diff(inh2sim(sAC,sCA),sAC);

$$\frac{1}{\left(\frac{1}{sAC}+\frac{1}{sCA}-1\right)^2 sAC^2}$$

diff(inh2sim(sAC,sCA),sCA);

$$\frac{1}{\left(\frac{1}{sAC}+\frac{1}{sCA}-1\right)^2 sCA^2}$$

inh2sim_nd := (sAC,sCA) -> ((1/((1/sAC+1/sCA-1)^2*sAC^2))^2 + (1/((1/sAC+1/sCA-1)^2*sCA^2))^2)^(1/2);

$$inh2sim\_nd := (sAC,sCA) \to \sqrt{\frac{1}{\left(\frac{1}{sAC}+\frac{1}{sCA}-1\right)^4 sAC^4}+\frac{1}{\left(\frac{1}{sAC}+\frac{1}{sCA}-1\right)^4 sCA^4}}$$

diff(sim2inh(sA,sB,simAB), sA);

$$-\frac{sB\,simAB}{sA^2(1+simAB)}$$

diff(sim2inh(sA,sB,simAB), sB);

$$\frac{simAB}{sA(1+simAB)}$$

diff(sim2inh(sB,sC,simBC), simBC);

$$\frac{1+\frac{sC}{sB}}{1+simBC}-\frac{\left(1+\frac{sC}{sB}\right)simBC}{(1+simBC)^2}$$

sim2inh_nd := (sB,sC,simBC) -> ((simAB/(1+simBC))^2 + (sB/sC^2*simBC/(1+simBC))^2 + ((1+sB/sC)/(1+simBC)-(1+sB/sC)*simBC/(1+simBC)^2)^2)^(1/2) *(Heaviside(simBC-max(((sB+sC-1)),0))-Heaviside(simBC-min(sB/sC,(sC/sB))));

$$sim2inh\_nd := (sB, sC, simBC) \rightarrow$$

$$\sqrt{\frac{simBC^2}{(1+simBC)^2} + \frac{sB^2\, simBC^2}{sC^4(1+simBC)^2} + \left(\frac{1+\frac{sB}{sC}}{1+simBC} - \frac{\left(1+\frac{sB}{sC}\right)simBC}{(1+simBC)^2}\right)^2}$$

$$\left(\text{Heaviside}(\text{simBC} - \max(\text{sB} + \text{sC} - 1, 0)) - \text{Heaviside}\left(\text{simBC} - \min\left(\frac{\text{sB}}{\text{sC}}, \frac{sC}{sB}\right)\right)\right)$$

We then produce the following diagrams:

plot3d(inh2sim_nd(sAC, sCA), sAC=0..1, sCA=0..1, axes=BOXED);

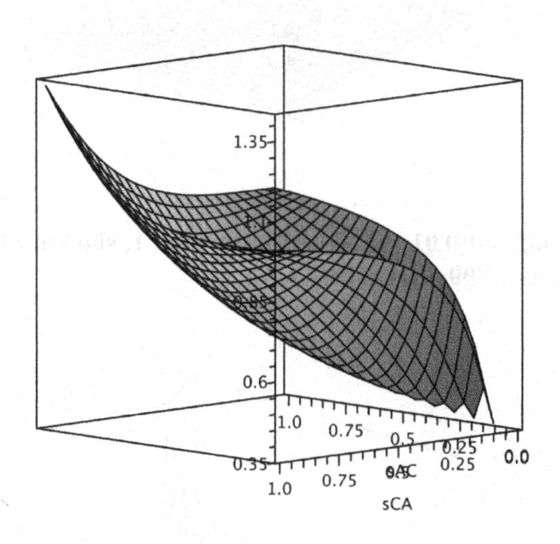

plot3d(sim2inh_nd(.1, sB, simAB), sB=0..1, simAB=0..1, resolution=400, axes=BOXED);

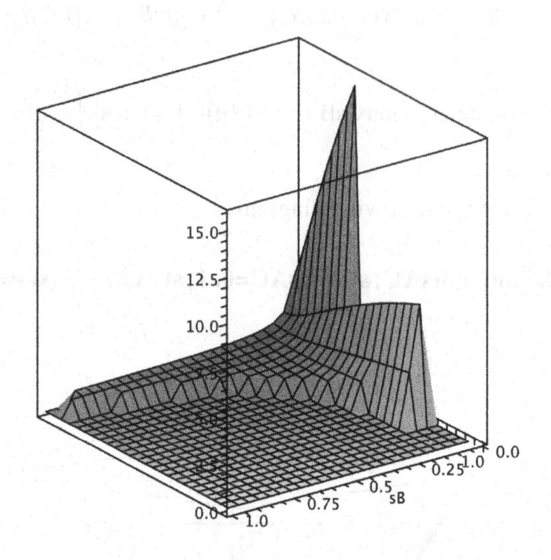

plot3d(sim2inh_nd(0.01, sB, simAB), sB = 0 .. 1, simAB = 0 .. 1, resolution = 800, numpoints = 800, axes = BOXED);

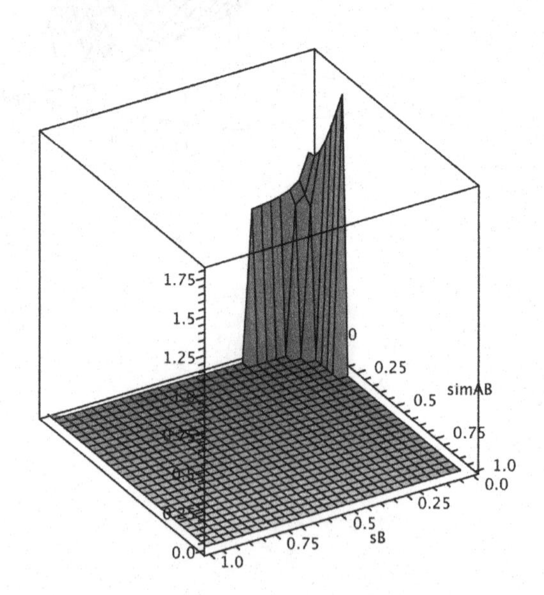

plot3d(sim2inh_nd(.4, sB, simAB), sB=0..1, simAB=0..1, resolution=800, axes=BOXED);

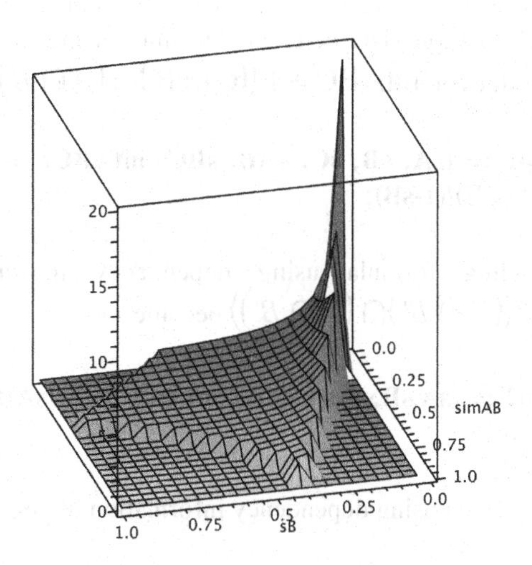

plot3d(sim2inh_nd(.9, sB, simAB), sB=0..1, simAB=0..1, axes=BOXED, resolution=800);

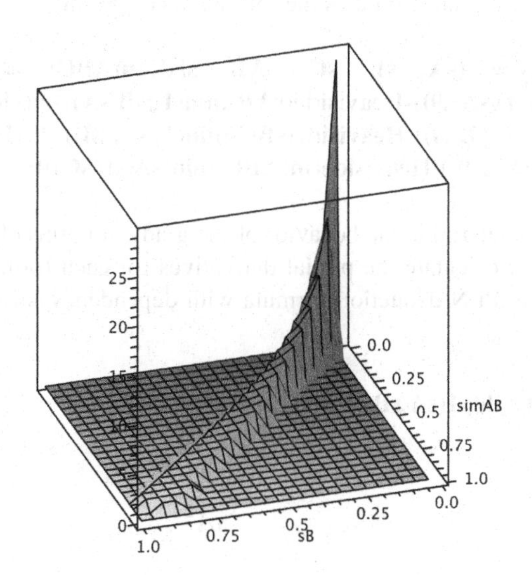

Depictions of the Gradient Norms of the Deduction Formulas Incorporating Dependency Information

Now we repeat the above exercise for the deduction variants that incorporate dependency information. The deduction formula using dependency information with a known value for $\text{intABC} = P((C \cap B) \cap (A \cap B))$ is:

dedAC_dep1 := (sA, sB, sC, sAB, sBC, intABC) -> intABC / sA + (1-sAB)*(sC-sB * sBC)/(1-sB);

The corresponding formula using dependency information with known $\text{intAnBC} = P((C \cap B') \cap (A \cap B'))$ becomes:

dedAC_dep2 := (sA,sB,sC,sAB,sBC,intAnBC) -> (sAB * sBC + intAnBC / sA);

Similarly, the formula using dependency information in both arguments is

dedAC_dep12 := (sA, intABC,intAnBC) -> intABC / sA + intAnBC / sA ;

For each of these formulas there are consistency constraints similar to those given earlier. Now we need to include an additional constraint on the intABC and/or the intAnBC variables. For dependency with a known value for intABC, for example, consistency implies the following constraint:

consistency:= (sA, sB, sC, sAB, sBC,intABC) -> (Heaviside(sAB-max(((sA+sB-1)/sA),0))-Heaviside(sAB-min(1,(sB/sA))))*(Heaviside(sBC-max(((sB+sC-1)/sB),0))-Heaviside(sBC-min(1,(sC/sB))))*(Heaviside(intABC-max(sA+sB+sC-2,0))-Heaviside(intABC-min(sA,sB,sC)));

We wish to determine the behavior of the gradient norms of these formulas. As before, we first calculate the partial derivatives for each formula. The partial derivatives of the PLN deduction formula with dependency information in the first argument are

diff(dedAC_dep1(sA,sB,sC,sAB,sBC,intABC),sA);

$$-\frac{\text{int}\, ABC}{sA^2}$$

diff(dedAC_dep1(sA,sB,sC,sAB,sBC,intABC),sB);

$$-\frac{(1-sAB)\,sBC}{1-sB}+\frac{(1-sAB)(sC-sB\,sBC)}{(1-sB)^2}$$

diff(dedAC_dep1(sA,sB,sC,sAB,sBC,intABC),sC);

$$\frac{1-sAB}{1-sB}$$

diff(dedAC_dep1(sA,sB,sC,sAB,sBC,intABC),sAB);

$$-\frac{sC-sB\,sBC}{1-sB}$$

diff(dedAC_dep1(sA,sB,sC,sAB,sBC,intABC),sBC);

$$-\frac{(1-sAB)sB}{1-sB}$$

diff(dedAC_dep1(sA,sB,sC,sAB,sBC,intABC),intABC);

$$\frac{1}{sA}$$

The norm of the gradient vector of the PLN deduction formula with dependency in the first argument is then:.

dedAC_nd_dep1 := (sA, sB,sC,sAB,sBC,intABC)->(((1-sAB)/(1-sB))^2+(1/sA)^2+(-(sC-sB*sBC)/(1-sB))^2+(-(1-sAB)*sB/(1-sB))^2+(-intABC/sA^2)^2+(-(1-sAB)*sBC/(1-sB)+(1-sAB)*(sC-sB*sBC)/(1-sB)^2)^2)^(1/2);

$$dedAC\_nd\_dep1 := (sA,sB,sC,sAB,sBC,intABC) \to \left(\frac{(1-sAB)^2}{(1-sB)^2}+\frac{1}{sA^2}+\frac{(sC-sB\,sBC)^2}{(1-sB)^2}+\frac{(1-sAB)^2 sB^2}{(1-sB)^2}+\frac{intABC^2}{sA^4}+\left(-\frac{(1-sAB)sBC}{1-sB}+\frac{(1-sAB)(sC-sB\,sBC)}{(1-sB)^2}\right)^2\right)^{1/2}$$

Similarly, the partial derivatives of the PLN deduction formulas with dependency information in the second argument are:

diff(dedAC_dep2(sA,sB,sC,sAB,sBC,intAnBC),sA);

$$-\frac{intABC}{sA^2}$$

diff(dedAC_dep2(sA,sB,sC,sAB,sBC,intAnBC),sB);

$$0$$

diff(dedAC_dep2(sA,sB,sC,sAB,sBC,intAnBC),sC);

$$0$$

diff(dedAC_dep2(sA,sB,sC,sAB,sBC,intAnBC),sAB);

$$sBC$$

diff(dedAC_dep2(sA,sB,sC,sAB,sBC,intAnBC),sBC);

$$sAB$$

diff(dedAC_dep2(sA,sB,sC,sAB,sBC,intAnBC),intAnBC);

$$\frac{1}{sA}$$

The norm of the gradient vector of the PLN deduction formula with dependency in the second argument then becomes:

dedAC_dep2_nd:=(sA,sB,sC,sAB,sBC,intAnBC) -> ((intAnBC/sA^2)^2 + (sBC)^2 +(sAB)^2+(1/sA)^2)^(1/2);

$$dedAC\_dep2\_nd := (sA,sB,sC,sAB,sBC,intAnBC) \to \sqrt{\frac{intAnBC^2}{sA^4} + sBC^2 + sAB^2 + \frac{1}{sA^2}}$$

Last, we calculate the norm of the gradient vector for the PLN deduction formula knowing the values for both intABC and intAnBC. The partial derivatives for this case are found using the following calculations.

diff(dedAC_dep12(sA,sB,sC, intABC,intAnBC),sA);

$$-\frac{intABC}{sA^2}-\frac{intAnBC}{sA^2}$$

diff(dedAC_dep12(sA,sB,sC,intABC,intAnBC),sB);

$$0$$

diff(dedAC_dep12(sA,sB,sC,intABC,intAnBC),sC);

$$0$$

diff(dedAC_dep12(sA,sB,sC,intABC,intAnBC),intABC);

$$\frac{1}{sA}$$

diff(dedAC_dep12(sA,sB,sC,intABC,intAnBC),intAnBC);

$$\frac{1}{sA}$$

Hence the norm of the gradient vector for PLN deduction when both intABC and intAnBC are known is:

dedAC_dep12_nd:=(sA,sB,sC,intABC,intAnBC)->((-intABC/sA^2-intAnBC/sA^2)^2+(1/sA)^2+(1/sA)^2)^(1/2);

$$dedAC\_dep12\_nd := (sA,sB,sC,intABC,intAnBC) \to \sqrt{\left(-\frac{intABC}{sA^2}-\frac{intAnBC}{sA^2}\right)^2+\frac{2}{sA^2}}$$

We now give graphical depictions of the gradient norms for the inference formulas that incorporate dependency. We first exhibit several representative samples when intABC is known.

plot3d(dedAC_nd_dep1(.1,.1,.5,.9,sBC,intABC)*consistency(.1,.1,.5,.9,sBC, intABC), sBC=0.. 1,intABC=0.. 1, axes=BOXED, labels=[sBC,intABC,dedAC_1_nd], numpoints=2000, resolution=800);

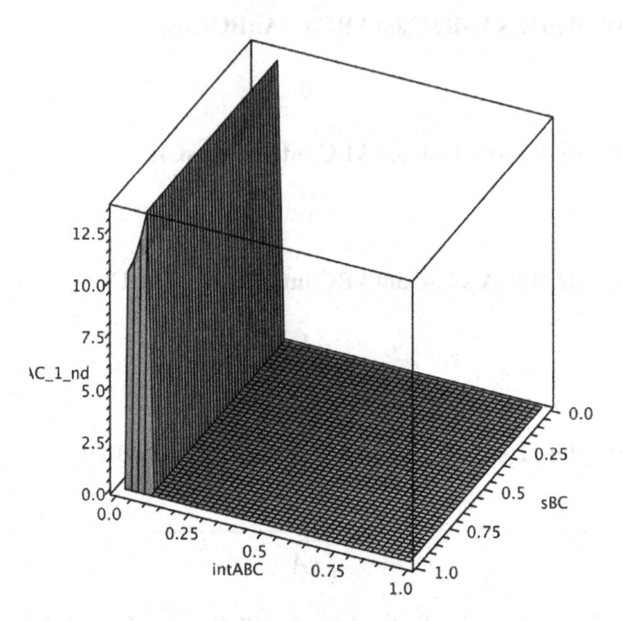

plot3d(dedAC_nd_dep1(.5,.5,.1,.5,sBC,intABC)*consistency(.5,.5,.1,.5,sBC, intABC), sBC=0.. 1,intABC=0.. 1, axes=BOXED, labels=[sBC,intABC,dedAC_1_nd], numpoints=1000, resolution=800);

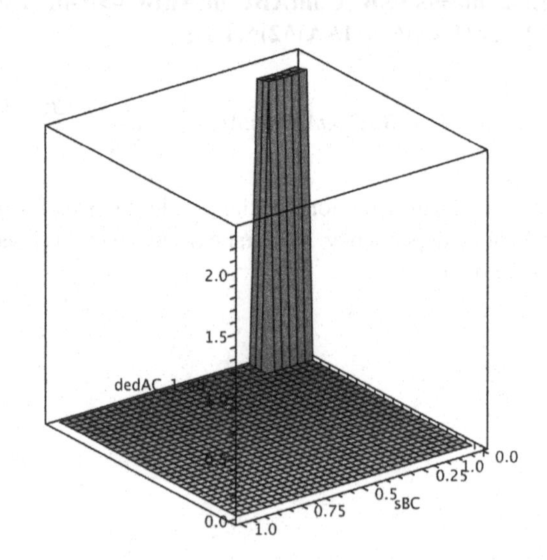

plot3d(dedAC_nd_dep1(.9,.5,.9,.5,sBC,intABC)*consistency(.9,.5,.5,.5,sBC, intABC), sBC=0.. 1,intABC=0.. 1, axes=BOXED, labels=[sBC,intABC,dedAC_1_nd], numpoints=2000, resolution=800);

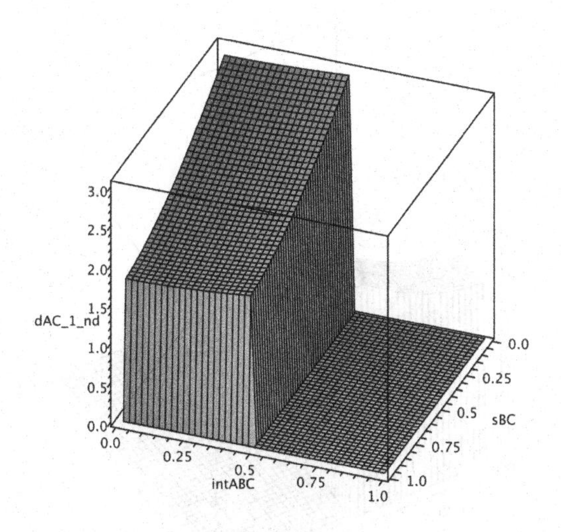

The next few graphs are representative of the deduction rule with a known value for intAnBC.

plot3d(dedAC_dep2_nd(.1,.1,.1,.5,sBC,intAnBC)*consistency(.1,.1,.1,.5,sBC,intAnBC), sBC=0.. 1,intAnBC=0.. 1, axes=BOXED, labels=[sBC,intAnBC,dedAC_2_nd], numpoints=800);

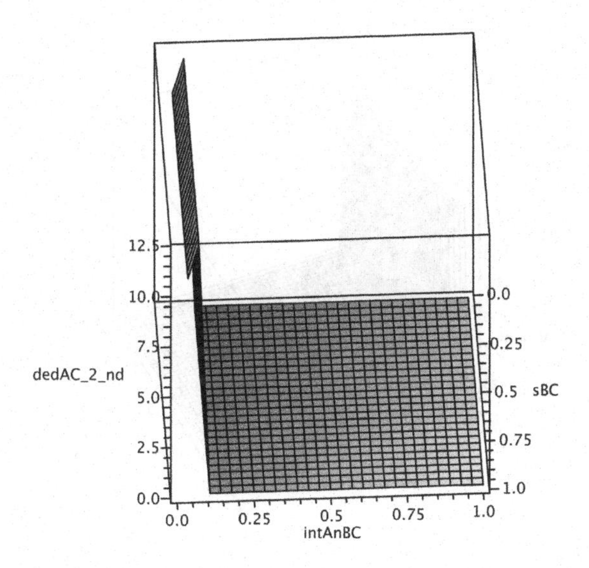

plot3d(dedAC_dep2_nd(.5,.5,.9,.5,sBC,intAnBC)*consistency(.5,.5,.9,.5,sBC,intAnBC), sBC=0.. 1,intAnBC=0.. 1, axes=BOXED, labels=[sBC,intAnBC,dedAC_2_nd], numpoints=2000, resolution=800);

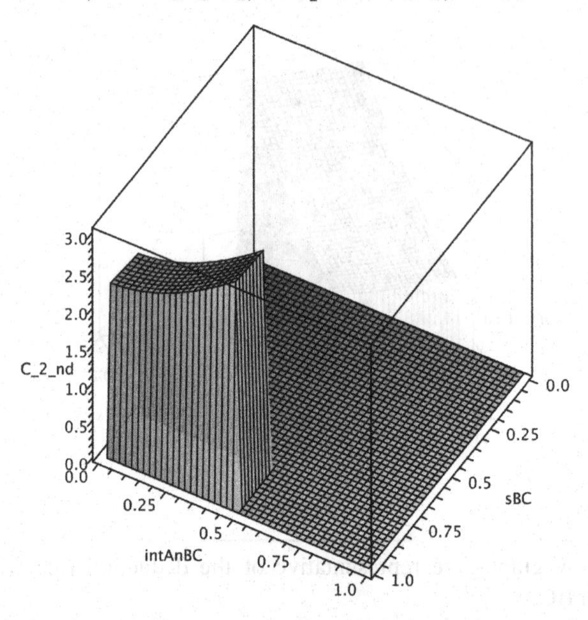

plot3d(dedAC_dep2_nd(.9,.1,.9,.1,sBC,intAnBC)*consistency(.9,.1,.9,.1,sBC,intAnBC), sBC=0.. 1,intAnBC=0.. 1, axes=BOXED, labels=[sBC,intAnBC,dedAC_2_nd], numpoints=2000, resolution=800);

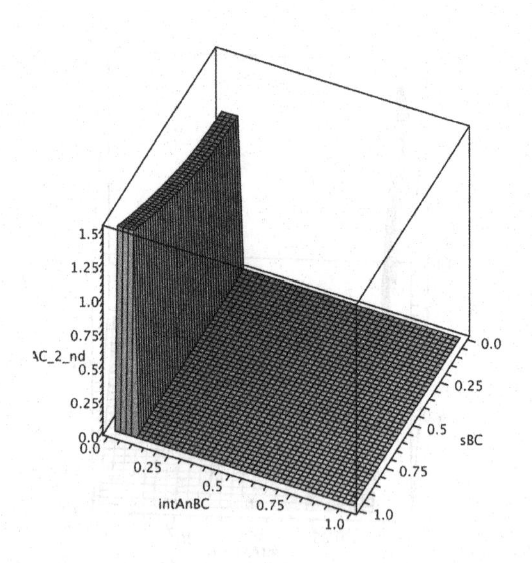

Finally we present a small sample of the graphs for deduction when both intABC and intAnBC are known.

plot3d(dedAC_dep12_nd(.1,.5,.9,intABC,intAnBC)*consistency(.1,.5,.9, .5,intABC, intAnBC), intABC=0.. 1, intAnBC=0.. 1, axes=BOXED, labels = [intABC, intAnBC, dedAC_12_nd], numpoints=2000, resolution=800);

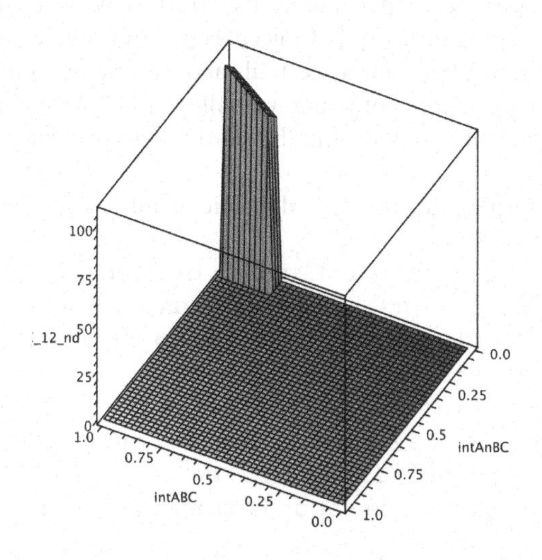

plot3d(dedAC_dep12_nd(.5,.5,.9,intABC,intAnBC) * consistency(.5,.5,.9, .5,intABC, intAnBC), intABC=0.. 1, intAnBC=0.. 1, axes=BOXED, labels = [intABC, intAnBC, dedAC_12_nd], numpoints=2000, resolution=800);

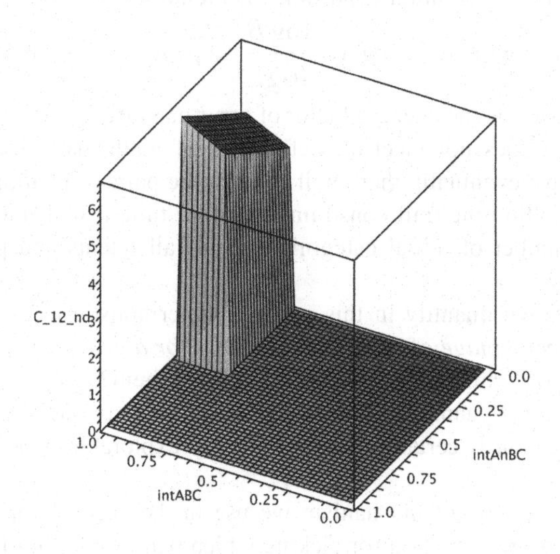

8.3 Formal Chaos in PLN

The above gradient norm results imply, among other things, the possibility of deterministic chaos associated with PLN. To see this, let's begin by constructing an artificially simple PLN-based reasoning dynamic.

Suppose one has a set of J entities, divided into M categories (represented by M Concept terms). Suppose one builds the $M(M-1)$ possible (probabilistically weighted) Subsets between these M ConceptTerms. Assume each of the Subsets maintains an arbitrarily long inference trail, and assume the Subsets are ordered overall (e.g., in an array). This ordering naturally defines an ordering on the set of pairs of Subsets as well. Let V denote the set of ConceptTerms and Subsets thus constructed.

Then one can define a deterministic dynamic on this space V, as follows:

```
1) Choose the first pair of Subsets that can be
   combined (using deduction, induction, or ab-
   duction) without violating the inference trail
   constraints.
2) Do the inference on the pair of Subsets.
3) Use the revision rule to merge the resulting
   new Subset in with the existing Subset that
   shares source and target with the newly cre-
   ated Subset.
```

The dynamics, in this scheme, depend on the ordering. But it is clear that there will be some orderings that result in a long series of error-magnifying inferences. For these orderings one will have genuinely chaotic trajectories, in the sense of exponential sensitivity to initial conditions. If the initial conditions are given to B bits of precision, it will take roughly $\frac{\log B}{\log z}$ steps for the inferred values to lose all meaning, where z is the average factor of error magnification.

This dynamic does not meet all definitions of mathematical chaos, because eventually it stops: eventually there will be no more pairs of relationships to combine without violating the trail constraints. But the time it will run is hyperexponential in the number of initial categories, so for all intents and purposes it behaves chaotically.

The big unknown quantity in this example deterministic inferential dynamic, however, is the *percentage of chaotic trajectories for a given set of initial categories*. How many orderings give you chaos? Most of them? Only a rare few? Half? The mathematics of dynamical systems theory doesn't give us any easy answers, although the question is certainly mathematically explorable with sufficient time and effort.

In the inference control mechanism we use in the current implementation of PLN, we use a different method for picking which pair of relationships to combine at a given step, but the ultimate effect is the same as in this toy deterministic inferential dynamic. We have a stochastic system, but we may then have stochastic chaos. The question is still how common the chaotic trajectories will be.

8.3.1 Staying on the Correct Side of the Edge of Chaos

We can see a very real meaning to the "Edge of Chaos" concept here (Langton 1991; Packard 1988). Doing reasoning in the error-magnification domain gives rise to chaotic dynamics, which is bad in this context because it replaces our strength values with bogus random bits. But we want to get as close to the error-magnification domain as possible – and dip into it as often as possible without destroying the quality of our overall inferences – so that we can do reasoning in as broadly inclusive a manner as possible.

Looking at the problem very directly, it would seem there are four basic approaches one might take to controlling error magnification:

Safe approach

Check, at each inference step, whether it's an error-magnifying step or not. If it's an error-magnifying step, don't do it.

Reckless approach

Just do all inference steps regardless of their error magnification properties.

Cautious approach

Check, at each inference step, whether it's an error-magnifying step or not. If the gradient-norm is greater than some threshold T (say, T=2 or T=1.5), then don't do it. The threshold may possibly be set differently for the different inference rules (some reasons to do this will be mentioned below).

Adaptive approach

Check, at each inference step, whether it's an error-magnifying step or not. If it's an error-magnifying step, decide to do it based on a calculation. Each logical Atom must keep a number D defined by

```
Initially, D = 1
At each inference step involving premises P1,…,Pn, set
Dnew = D * (norm of the gradient of the pertinent inference rule evaluated at (P1 … Pn) )
```

```
In the adaptive approach, the decision of whether to do an inference step or not is guided by whether it can be done while keeping D below 1 or not.
```

The reckless approach is not really viable. It might work acceptably in a deduction-only scenario, but it would clearly lead to very bad problems with inversion, where the derivative can be huge. We could lose three or four bits of precision in the truth-value in one inference step. The adaptive approach, on the other hand, is very simple code-wise, and in a way is simpler than the cautious approach, because it involves no parameters.

In practice, in our current implementation of PLN we have not adopted any of these approaches because we have subsumed the issue of avoiding error magnification into the larger problem of weight-of-evidence estimation. The adaptive approach to avoiding error magnification winds up occurring as a consequence of an adaptive approach to weight-of-evidence-based inference control.

Conceptually speaking, we have reached the conclusion that ***speculative reasoning, unless carefully controlled, or interwoven with larger amounts of non-speculative reasoning, can lead to significant error magnification***. Sometimes a cognitive system will want to do speculative error-magnifying reasoning in spite of this fact. But in these cases it is critical to use appropriately adaptive inference control. Otherwise, iterated speculative inference may lead to nonsense conclusions after a long enough chain of inferences – and sometimes after very short ones, particularly where inversion is involved.

Chapter 9: Large-Scale Inference Strategies

Abstract In chapter 9, we consider special inference approaches useful for inference on large bodies of knowledge.

9.1 Introduction

The inference rules presented in the previous chapters follow a common format: given a small number of premises (usually but not always two), derive a certain conclusion. This is the standard way logical reasoning is described, yet it is not the only way for logical reasoning to be carried out in real inference systems. Another equally important application of logical inference is to derive conclusions integrating large bodies of information. In these cases the standard logical rules are still applicable, but thinking about reasoning in terms of the step-by-step incremental application of standard logical rules is not necessarily the best approach. If one considers "large-scale inference" (or alternatively, "holistic inference") as a domain unto itself, one quickly arrives at the conclusion that there are different control strategies and even in a sense different inference rules appropriate for this situation. This chapter presents four such specialized approaches to large-scale inference:

- Batch inference, which is an alternative to the Rule of Choice usable when one has to make a large number of coordinated choices based on the same set of data
- Multideduction, which is an alternative to iterated deduction and revision that is appropriate when one has to do a series of deductions and revision based on the same large body of data
- Use of Bayesian networks to augment and accelerate PLN inference
- Trail-free inference, the most radical alternative approach considered here, in which trails are omitted and inference is considered as a nonlinear multivariable optimization problem

No doubt there are other viable approaches besides these ones. In a larger sense, the point of this chapter is to indicate the diversity of approaches to deriv-

nclusions from probabilistic premises, beyond traditional incremental logical theorem proving type approaches.

9.2 Batch Inference

In this section we return to the theme of handling circular inference. The Rule of Choice options mentioned in Chapter 5 are relatively simple and have served us adequately in experiments with PLN. However, such an approach is clearly not optimal in general – it's basically a "greedy" strategy that seeks good inference trajectories via local optimizations rather than by globally searching for good inference trajectories. In a large-scale inference context, it is possible to work around this problem by carrying out a large number of inference steps en masse, and coordinating these steps intelligently.

This kind of "batch" oriented approach to Rule of Choice type operations is probably not suitable for real-time reasoning or for cognitive processes in which reasoning is intricately interlaced with other cognitive processes. What it's good for is efficiently carrying out a large amount of reasoning on a relatively static body of information. The underlying motivation is the fact that, in many cases, it is feasible for a reasoning system to operate by extracting a large set of logical relationships from its memory, doing a lot of inferences on them at once, and then saving the results.

The method described in this section, "batch reasoning graph inference" (BRGI), tries to generate the maximum amount of information from a *set of inferences* on a body of data S, by generating an optimal set of inference trails for inferences from S. This process minimizes the use of the Rule of Choice. The BRGI process carries out choices similar to what the Rule of Choice does in searching for the optimal tree, but in a significantly more efficient way due to its more global scope. In a sense, it makes a lot of "Rule of Choice" type choices all at once, explicitly seeking a globally valuable combination of choices.

9.2.1 Batch Reasoning Graphs

Suppose we have a collection $S = \{S_1, \ldots, S_n\}$ of terms or logical relationships, drawn from a reasoning system's memory at a single point in time. Define a "batch reasoning graph" (brg) over S as a DAG (directed acyclic graph) constructed as follows:

- The leaf terms are elements of S.
- All non-leaf terms have two children.
- Each non-leaf term is labeled with a logical conclusion, which is derived from its children using one of the PLN inference rules.

In such a graph, each term represents a relationship that is derived directly from its children, and is derived indirectly from its grandchildren and more distant progeny.

When a brg term *A* is entered into the memory as a relationship, it should be given a trail consisting of its progeny in the tree. If we wish to conserve memory, we may form a smaller trail considering only progeny down to some particular depth.

For two terms *A* and *B*, if there is no path through the brg from *A* to *B* (following the arrows of the graph), then *A* and *B* may be considered "roughly independent," and can be used together in inferences. Otherwise, *A* and *B* are based on the same information and shouldn't be used together in inference without invocation of the Rule of Choice.

From any given set *S* it is possible to construct a huge variety of possible batch reasoning graphs. The batch reasoning process proposed here defines a quality measure, and then runs a search process aimed at finding the highest-quality brg. Many search processes are possible, of course. One option is to use evolutionary programming; crossing over and mutating DAGs is slightly tricky, but not as problematic as crossing over and mutating cyclic graphs. Another option is to use a greedy search algorithm, analogously to what is often done in Bayes net learning (Heckerman 1996).

What is the quality measure? The appropriate measure seems to be a simple one. First, we must define the expectation of an inference. Suppose we have an inference whose conclusion is an Atom of type L, and whose truth-value involves (s,d) = (strength, weight of evidence). This expectation of this inference is

$$d * p(s, L)$$

where $p(s, L)$ is the probability that a randomly chosen Atom of type *L*, drawn from the knowledge-store of the system doing the inference, has strength *s*. A crude approximation to this, which we have frequently used in the past, is

$$d * |s - 0.5|$$

So, suppose one creates a brg involving *N* inferences. Then one may calculate *T*, the total *expectation* of the set of inferences embodied in the brg. The ratio *T*/*N* is a measure of the average information per relationship in the brg. A decent quality measure, then, is

$$c \cdot N + (1 - c)\frac{T}{N}$$

eans that we want to do a lot of inferences, but we want these inferences to be high quality.

This concept, as specifically described above, is directly applicable both to FOI and to all parts of HOI except those that involve Predicates. In order to apply it to HOI on Predicates, the definition of a brg has to be extended somewhat. One has to allow special brg terms that include logical operators: AND, OR, or NOT. A NOT term has one input; AND and OR terms may be implemented with two inputs, or with *k* inputs. In assessing the quality of a brg, and counting the expectations of the terms, one skips over the relationships going into these logical operator terms; the value delivered by the operator terms is implicit in the expectations of the relationships coming out of them (i.e., the inferences that are made from them).

9.2.2 Example of Alternate Batch Reasoning Graphs

To make these ideas more concrete, let us consider some simple examples. Suppose the system contains the relations

```
Inheritance A B

Inheritance B C

Inheritance C A
```

with strengths derived, not through reasoning, but through direct examination of data.

These examples may lead to either of the following graphs (trees), where Inh is short for Inheritance:

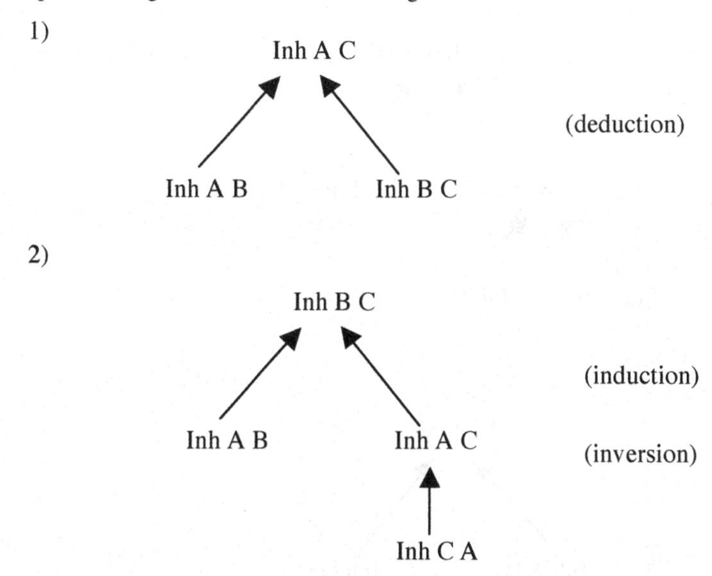

In Graph 1 (Inh B C) is used to derive (Inh A C), whereas in Graph 2 (Inh A C) is used to derive (Inh B C). So once you've done the reasoning recorded in Graph 2, you cannot use (Inh B C) to increase your degree of belief in (Inh A C), so you cannot use the reasoning recorded in Graph 1. And once you've done the reasoning recorded in Graph 1, you cannot use (Inh A C) to increase your degree of belief in (Inh B C), so you cannot do the reasoning involved in Graph 2.

Which of these is a better graph, 1 or 2? This depends on the truth-values of the relationships at the terms of the brg. If (Inh C A) has a much higher empirically derived strength than (Inh B C), then it may make sense to construct the inference given in graph 2; this may produce a higher-total-average-expectation group of relationships.

For a slightly more complex example, consider the following two graphs:

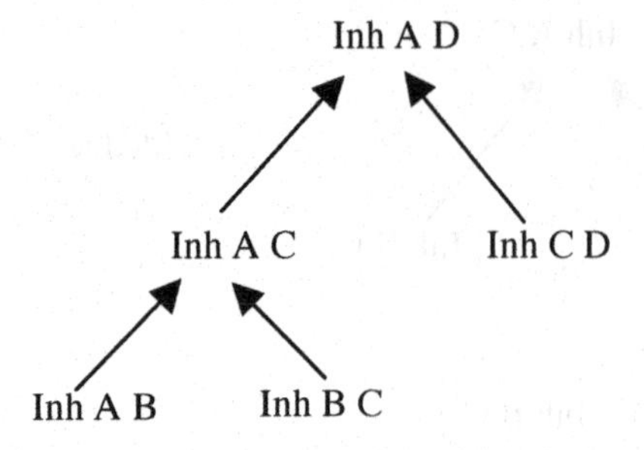

2)

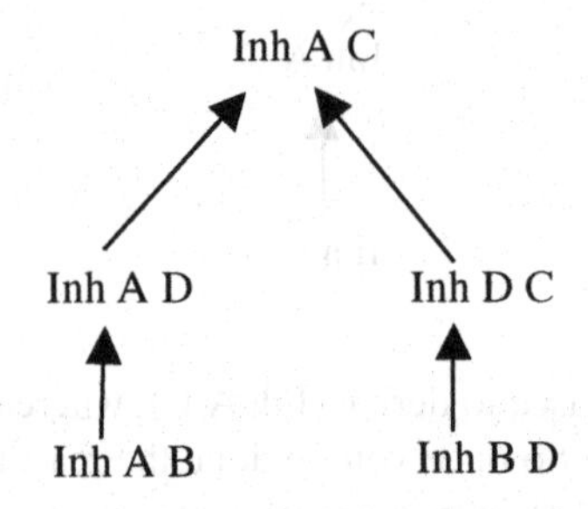

In Case 1, the inference

```
Inh A D
Inh A B
|-
Sim B D
```

cannot be drawn, because (Inh A B) is in the trail of (Inh A D).
In Case 2, on the other hand, the inference

```
Inh A C
Inh A B
|-
Sim B C
```

cannot be drawn because (Inh A B) is in the trail of (Inh A C).
So the above trees can be expanded into directed acyclic graphs of the form

1)

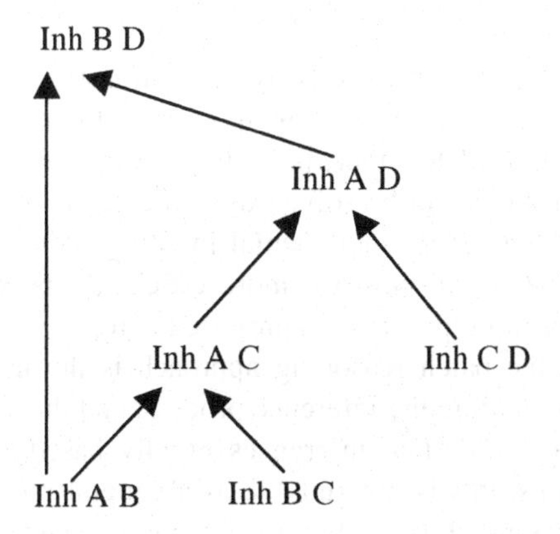

2)

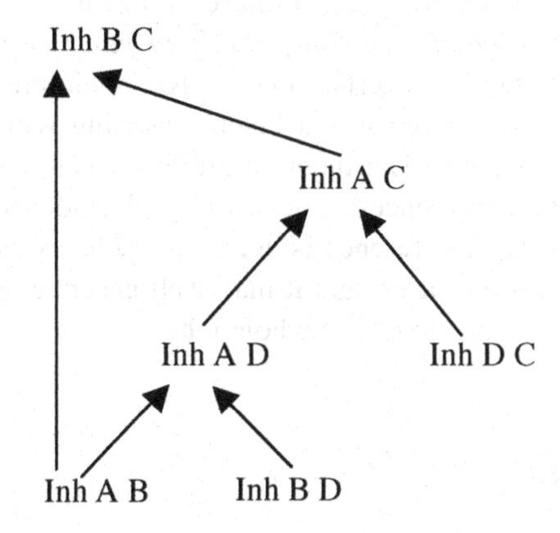

Which is a better path for reasoning to follow? This depends on the truth-values of the relationships involved. In particular, suppose that

```
(Inheritance B D).TruthValue.expectation =
(Inheritance B C).TruthValue.expectation
```

prior to any of these reasoning steps being carried out. Then the quality of 1 versus 2 depends on the relative magnitudes of

```
(Inheritance D C).TruthValue.expectation
```

and

```
(Inheritance C D).TruthValue.expectation
```

Discussion

In doing inference based on a large body of relationships, the sort of decision seen in the above simple examples comes up over and over again, and multiple decisions are relationshipped together, providing an optimization problem. This sort of optimization problem can be solved dynamically and adaptively inside an inference system by the standard incremental inference control mechanisms and the Rule of Choice, but it can be done more efficiently by considering it as a "batch inference optimization process" as proposed here.

What one loses in the batch reasoning approach is the involvement of non-inferential dynamics in guiding the inference process, and the ability to carry out a small number of highly important inferences rapidly based on a pool of background knowledge. Thus, this is not suitable as a general purpose inference system's only inference control strategy, but it is a valuable inference control strategy to have, as one among a suite of inference control strategies.

Most particularly, the strength of batch inference lies in *drawing a lot of relatively shallow inferences based on a large body of fairly confident knowledge.* Where there is primarily highly uncertain knowledge, noninferential methods will be so essential for guiding inference that batch reasoning will be inappropriate. Similarly, in cases like mathematical theorem-proving where the need is for long chains of inference ("deep inferences"), noninferential guidance of the reasoning process is also key, and batch inference is not going to be as useful. Batch inference can still be used in these cases, and it may well generate useful information, but it will definitely not be able to do the whole job.

9.3 Multideduction

Next, we present a complementary approach to large-scale inference that is more radical than batch inference in that it proposes not just a special approach to inference control oriented toward large-scale inference, but a special way of calculating truth-values under the assumption that there is a large amount of evidence to be integrated simultaneously. We introduce a kind of amalgamated inference rule called "multideduction," which carries out deduction and revision all at once across a large body of relationships, and as a consequence avoids a lot of the compounding of errors that occurs as a result of doing repeated deductions and revisions.

In brief, what multideduction does is to perform the inference

$$\begin{array}{l} A \rightarrow \{B_1, B_2, B_3, \cdots B_k\} \\ \underline{\{B_1, B_2, B_3, \cdots B_k\} \rightarrow C} \\ \therefore A \rightarrow C \end{array}$$

in a single step.

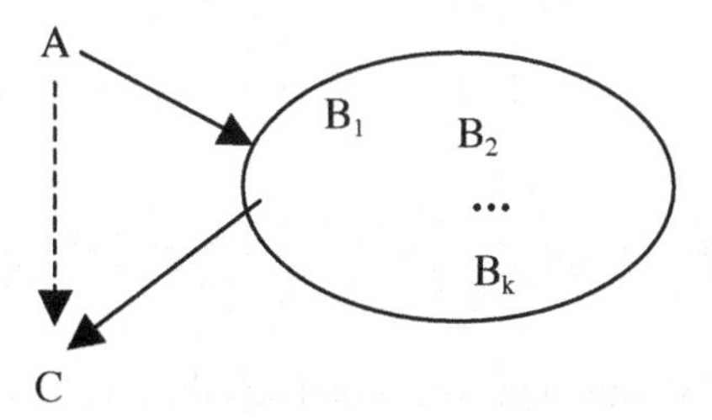

Multideduction

This is a change from the standard approach, in which one does such inference for each node B_i separately, using the deduction rule, and then merges these results using the revision rule.

Multideduction, as described here, incorporates the same independence assumptions about each of the nodes, B_i, used in the standard deduction formula, and uses some approximation to do its "revision"; but we believe these approximations are less erroneous than the ones used in the standard "repeated pair-wise deduction and revision" approach. Instead, multideduction is equivalent to "repeated pair-wise deduction and revision" where revision is done using a more sophisticated revision rule customized for revision of two deductive conclusions (a rule that involves subtracting off for overlap among the nodes of the two deductions). There is also a variant of multideduction that is based on the concept geometry based deduction formula rather than the independence assumption based deduction formula; this variant is not explicitly described here but can be easily obtained as a modification of the independence based multideduction formulation.

Multideduction works naturally with an inference control strategy in which one takes the following steps in each "inference cycle":

- Do node probability inference on all sufficiently important nodes.
- Do inversion on all sufficiently important relationships.
- For all sufficiently important relationships (InheritanceRelationship A C), calculate s_{AC} from all B-nodes for which s_{AB} and s_{BC} are available.

The gauging of importance is beyond the scope of this section (and is not purely a PLN matter, but gets into non-inferential issues regarding the framework into which PLN is integrated).

9.3.1 The Inclusion-Exclusion Approximation Factor

The inclusion-exclusion formula from standard combinatorics can be written as

$$P\left(\bigcup_{i=1}^{n} S_i\right) = \sum_{i=1}^{n} P(S_i) - \sum_{i,j=1}^{n} P(S_i \cap S_j) + \text{error} = \text{term1} + \text{term2} + \text{error}.$$

We may estimate the expected error of using only the first two terms via a formula

$$\text{mean}(\text{error}) = \beta\left(\frac{\text{term1}}{n}, \frac{\text{term2}}{n^2}\right)$$

where β is an "inclusion-exclusion approximation factor." Points on the graph of β may be computed using a simulation of random sets or random rectangles, or they may be computed in a domain-specific manner (using example sets S_i of interest). The values for β may be pre-computed and stored in a table. More sophisticatedly, it seems one might also be able to make a formula

$$\text{mean}(\text{error}) = \beta\left(P(S_i), P(S_i \cap S_j)\right)$$

by doing some complicated summation mathematics; or use a combination of summation mathematics with the simulation-estimation approach. Generally, the idea is that we have no choice but to use the first two terms of the inclusion-exclusion formula as an approximation, because we lack the information about higher-order intersections that the other terms require and because the other terms are too numerous for pragmatic computation; but we can make a reasonable effort to estimate the error that this assumption typically incurs in a given domain. So we will use the formula

$$P\left(\bigcup_{i=1}^{n} S_i\right) = \sum_{i=1}^{n} P(S_i) - \sum_{i,j=1}^{n} P(S_i \cap S_j) + \beta$$

where the β will be written without arguments, for compactness, but is intended to be interpreted in accordance with the above comments. Note that this is an additive "fudge factor" and that it must be used with a "stupidity check" that avoids assigning $\left|\bigcup_{i=1}^{n} S_i\right|$ a value greater than 1.

9.3.2 Multideduction Strength Formula

We now move on to the actual multideduction formula. We assume we have two nodes *A* and *C*, and a set of nodes $\{B_i\}$ for which we know s_A, s_C s_{Bi}, s_{ABi}, s_{BiC}, and s_{BiBj}. We wish to derive the value s_{AC}.

We write

$$B = \bigcup_{i=1}^{n} B_i$$

We will calculate the deduction truth-value via the formula

$$P(C|A) = \frac{P(A \cap C)}{P(A)} = \frac{P(A \cap C \cap B) + P(A \cap C \cap \neg B)}{P(A)}$$

$$= \frac{P(A \cap C \cap B)}{P(A)} + \frac{+P(A \cap C \cap \neg B)}{P(A)}$$

The key calculation is

$$P(A \cap C \cap B) = P\left(A \cap C \cap \bigcup_{i=1}^{n} B_i\right) + \beta_1 \mathrm{P(A \cap C \cap B)}$$

[now we assume A and C are independent in the B_i and the $B_i \cap B_j$]

$$= \sum_{i=1}^{n} \frac{(A \cap B_i)P(C \cap B_i)}{P(B_i)} -$$

$$\sum_{\substack{i,j=1,n \\ i \neq j}}^{n} \frac{P(A \cap B_i \cap B_j)P(C \cap B_i \cap B_j)}{P(B_i \cap B_j)} + \beta_1$$

[now we introduce the heuristic approximation

$$P(A \cap B_i \cap B_j) = P(B_i \cap B_j)\left(w_{AB_iB_j} P(A|B_i) + \left(1 - w_{AB_iB_j}\right)P(A|B_j)\right)$$

where

$$w_{AB_iB_j} = \frac{d_{B_iA}}{d_{B_iA} + d_{B_jA}}$$

(the d's are "weights of evidence" of the indicated relationships) and a similar approximation for $\mathrm{P}(A \cap B_j \cap B_i)$.

$$\sum_{i=1} \frac{P\ A \cap B_i)P(C \cap B_i)}{P(B_i)}$$

$$-\sum_{\substack{i=1 \\ i \neq j}}^{n} \left[\begin{array}{l} P(B_i \cap B_j)\left(w_{AB_iB_j} P(A|B_i) + \left(1 - w_{AB_iB_j}\right)P(A|B_j)\right) \\ \cdot\left(w_{CB_iB_j} P(C|B_i) + \left(1 - w_{CB_iB_j}\right)P(C|B_j)\right) \end{array} \right] + \beta_1$$

$$= \sum_{i=1}^{n} P(B_i|A)P(A)P(C|B_i)$$

$$-\sum_{\substack{i,j=1 \\ i \neq j}}^{n} \left[\begin{array}{l} P(B_j|B_i)P(B_i)\left(w_{AB_iB_j} \frac{P(B_i|A)P(A)}{P(B_i)}\right) \\ +\left(1 - w_{AB_iB_j}\right)\frac{P(B_j|A)P(A)}{P(B_j)}\left(w_{CB_iB_j} \frac{P(C|B_i)P(A)}{P(B_i)} + \left(1 - w_{CB_iB_j}\right)\frac{P(B_j|C)P(A)}{P(B_j)}\right) \end{array} \right] + \beta_1$$

Dividing both sides by $P(A)$ and simplifying a little yields

$$\frac{P(A \cap C \cap B)}{P(A)}$$

$$= \sum_{i=1}^{n} s_{AB_i} s_{B_iC} - \sum_{\substack{i.j=1 \\ i \neq j}}^{n} s_{B_iB_j} \left[\begin{array}{l} \left(w_{AB_iB_j} s_{AB_i} + \left(1 - w_{AB_iB_j}\right)\frac{s_{AB_j} s_{B_i}}{s_{B_j}}\right) \\ \cdot\left(w_{CB_iB_j} s_{CB_i} + \left(1 - w_{CB_iB_j}\right)\frac{s_{CB_j} s_{B_i}}{s_{B_j}}\right) \end{array} \right] + \beta_1$$

which should look immediately intuitively sensible. We have the sum of all the deduction first-terms for all the B_i, minus the sum of some terms proportional to the intersections between the B_i.

The second term in the multideduction rule may be calculated similarly; we have

$$P(A \cap C \cap \neg B) = P\left(A \cap C \cap \neg\left(\bigcup_{i=1}^{n} B_i\right)\right) = P\left(A \cap C \cap \left(\bigcup_{i=1}^{n} \neg B_i\right)\right)$$

and hence the calculation is the same as for the first term, but with $\neg B_i$ substituted for B_i everywhere. Hence we have

$$\frac{P(A \cap C \cap \neg B)}{P(A)}$$

$$= \sum_{i=1}^{n} s_{A\neg B_i} s_{\neg B_i C} - \sum_{\substack{i,j=1 \\ i \neq j}}^{n} s_{\neg B_i \neg B_j} \left[\begin{array}{l} \left(w_{A\neg B_i \neg B_j} s_{A\neg B_i} + \left(1 - w_{A\neg B_i \neg B_j}\right) \frac{s_{A\neg B_j} s_{\neg B_i}}{s_{\neg B_j}} \right) \\ \cdot \left(w_{C\neg B_i \neg B_j} s_{C\neg B_i} + \left(1 - w_{C\neg B_i \neg B_j}\right) \frac{s_{C\neg B_j} s_{\neg B_i}}{s_{\neg B_j}} \right) \end{array} \right] + \beta_2$$

to obtain the final multideduction formula:

$$\begin{aligned} s_{AC} &= \sum_{i=1}^{n} \left(s_{AB_i} s_{B_i C} + s_{A\neg B_i} s_{\neg B_i C} \right) \\ &- \sum_{\substack{i,j=1 \\ i \neq j}}^{n} s_{B_i B_j} \left(w_{AB_i B_j} s_{AB_i} + \left(1 - w_{AB_i B_j}\right) \frac{s_{AB_j} s_{B_i}}{s_{B_j}} \right) \left(w_{CB_i B_j} s_{CB_i} + \left(1 - w_{CB_i B_j}\right) \frac{s_{CB_j} s_{B_i}}{s_{B_j}} \right) \\ &- \sum_{\substack{i,j=1 \\ i \neq j}}^{n} s_{\neg B_i \neg B_j} \left(w_{A\neg B_i \neg B_j} s_{AB_i} + \left(1 - w_{A\neg B_i \neg B_j}\right) \frac{s_{A\neg B_j} s_{\neg B_i}}{s_{\neg B_j}} \right) \left(w_{C\neg B_i \neg B_j} s_{C\neg B_i} + \left(1 - w_{C\neg B_i \neg B_j}\right) \frac{s_{C\neg B_j} s_{\neg B_i}}{s_{\neg B_j}} \right) \\ &+ \beta_1 + \beta_2 \end{aligned}$$

which may be evaluated using the originally given probabilities using the substitutions

$$s_{A\neg B_i} = 1 - s_{AB_i}$$

$$s_{\neg B_i C} = \frac{s_C - s_{B_i C}}{1 - s_{B_i}}$$

$$s_{\neg B_i \neg B_j} = \frac{1 + s_{B_j} - s_{B_i B_j} s_{B_i}}{1 - s_{B_i}}$$

The weights

$$w_{A\neg B_i \neg B_j} = \frac{d_{\neg B_i A}}{d_{\neg B_i A} + d_{\neg B_j A}}$$

also require expansion. We may estimate

$$N_{\neg B_i A} = N_{\neg B_i} s_{B_i A} = \left(N_U - N_{B_i}\right) s_{\neg B_i A}$$

N_U is the universe size, and calculate the weight of evidence $d_{\neg BiA}$ from the count $N_{\neg BiA}$; and do similarly for $d_{\neg BjA}$.

Looking at this formula as a whole, what we have is the sum of the individual deductive conclusions, minus estimates of the overlap between the terms. Note how different this is from the standard deduction-plus-revision approach, which averages the individual deductive conclusions. Here the (weighted) averaging approximation is done inside the second term where the overlaps of *A* and *C* individually inside each $B_i \cap B_j$ and $\neg B_i \cap \neg B_j$ are estimated. Multideduction is similar to deduction-plus-revision where revision is done using an overlap-savvy form of the revision rule.

One could extend this formula in a obvious way to make use of triple-intersection information $P(B_i \cap B_j \cap B_k)$. This could be useful in cases where a significant amount of such information is available. However, the number of B_i nodes, *n*, must be quite small for this to be tractable because the third term of the inclusion-exclusion formula has $\binom{n}{3}$ terms.

9.3.3 Example Application of Multideduction

In this section we present an example application where multideduction adds significant value. The example is drawn from a software system that utilizes the NCE for controlling virtual animals in the Second Life virtual world. As described in (Goertzel 2008), that system as a whole has numerous aspects and is focused on learning of novel behaviors. The aspect we will discuss here, however, has to do with the generation of spontaneous behaviors and the modification of logical predicates representing animals' emotions.

The full rule-base used to guide spontaneous behaviors and emotions in this application is too large to present here, but we will give a few evocative examples. We begin with a few comments on rule notation. Firstly, the notation ==> in a rule indicates a PredictiveImplicationRelationship. Rules are assumed to have truth-value strengths drawn from a discrete set of values

```
0
VERY LOW
LOW
MIDDLE
HIGH
VERY HIGH
1
```

In the following list, all rules should be assumed to have a truth-value of HIGH unless something else is explicitly indicated.

For clarity, in the following list of rules we've used suffixes to depict c types of entities: P for personality traits, E for emotions, C for contexts, and S for schemata (the latter being the lingo for "executable procedures" within the NCE). In the case of schemata an additional shorthanding is in place; e.g., barkS is used as shorthand for (Execution bark) where bark is a SchemaNode. Also, the notation TEBlah($X) is shorthand for

```
ThereExists $X
   Evaluation Blah $X
```

i.e., an existential quantification relationship, such as will be discussed in Chapter 11.

Example rules from the rule-base are as follows:

```
angerToward($X) ==> angry
loveToward($X) ==> love
hateToward($X) ==> hate
fearToward($X) ==> fear
TEgratitudeToward($X) ==> gratitude
angerToward($X) ==> ¬friend($X) <LOW>
TE(near($X) & novelty($X)) ==> novelty
TEloveToward($X) & sleepy ==> gotoS($X)
TE(loveToward($X) & near($X)) & sleepy ==> sleepS
gratitudeToward($X) ==> lick($X)
atHomeC & sleepyB ==> Ex sleepS <.7>
gotoS($X) ==> near($X) <.6>
gotoS($X) ==> near($X) <.6>
AggressivenessP & angryE & barkS => happyE
AggressivenessP & angryE & barkS ==> proudE
AggressivenessP & angerToward($X) ==> barkAtS($X) <VERY HIGH>
AggressivenessP & angerToward($X) ==> barkAtS($X) <VERY HIGH>
AggressivenessP & angerToward($X) ==> nipS($X) <MIDDLE>
AggressivenessP & near($X) & ¬friend($X) ==> angerToward($X)
AggressivenessP & near($X) & enemy($X) ==>
                              angerToward($X) <VERY HIGH>
AggressivenessP & near_my_food($X) & ¬friend($X) ==>
                              angryToward($X) <VERY LOW>
AggressivenessP & near_my_food($X) ==> angryToward($X)
AggressivenessP & angerToward($X) & ¬friend($X) ==> hate($X)
AggressivenessP & OtherOrientationP & ownerNear($X) & enemy($X)
                               ==> angerToward($X)
AggressivenessP & near($X) & enemy($X) & homeC ==> angerToward($X)
AggressivenessP & ¬happyE & ¬angryE ==> boredE
AggressivenessP & jealousE ==> angryE
AggressivenessP & boredE ==> angryE <LOW>
```

Spontaneous activity of a virtual animal, governed by the above equations, is determined based on the modeling of habitual activity, such as the carrying out of actions that the pet has previously carried out in similar contexts. For each schema S, there is a certain number of implications pointing into (Ex S), and each of these implications leads to a certain value for the truth-value of (Ex S). These values may be merged together using (some version of) the revision rule.

However, a complication arises here, which is the appearance of emotion values like happyE on the rhs of some implications, and on the lhs of some others. This requires some simple backward chaining inference in order to evaluate some of the (Ex S) relationships (note that the details of the PLN backward chaining methodology will be described in Chapter 13). And this is an example where multideduction may be extremely valuable. Essentially, what we are doing here is carrying out revision of a potentially large set of deductive conclusions. We saw in our original discussion of revision for simple truth-values that in the case of revision of deductive conclusions, the standard revision formula can be improved upon considerably. Multideduction provides an even more sophisticated approach, which can naturally be extended to handle indefinite probabilities. Using multideduction to estimate the truth-value of (Ex S) for each schema S provides significantly reduced error as compared to the multiple-deductions-followed-by-revision methodology.

A similar approach applies to the generation of goal-driven activity based on rules such as the above. As an example, suppose we have a goal G that involves a single emotion/mood E, such as excitement. Then there are two steps:

1. Make a list of schemata S whose execution is known to fairly directly affect E.
2. For these schemata, estimate the probability of achievement of G if S were activated in the current context.

For Step 1 we can look for

- Schemata on the lhs of implications with E on the rhs
- One more level: schemata on the lhs of implications with X on the rhs, so that X is on the lhs of some implication with E on the rhs

In the case of a large and complex rule-base, Step 2 may be prohibitively slow without some sort of batch-inference based approach. But for moderate-sized rule-bases, a simple approach is unproblematic. However, there is the same problem as there is in the case of spontaneous activity: a lot of deductions followed by revisions, which produce considerable error that may be removed by taking a multideduction approach.

9.4 Using Bayes Nets to Augment PLN Inference

Pursuing further the example of the previous section, we now explain how Bayes nets techniques may be used, much in the manner of batch inference, to accelerate inference based on a knowledge base that is relatively static over time.

Keeping the example rule-base from the previous section in mind, suppose we have a target predicate P, and have a set S of implications I(k) of the form

```
A(k, 1) & A(k, 2) & ... & A(k, n(k)) ==> P <s_k>
```

where each A(k, i) is some predicate (e.g., an emotion, personality trait, or context), and obviously A(k, j)=A(m, i) sometimes. Then this set may be used to spawn a Bayes net that may be used to infer the truth-value of

```
A(k+1, 1) & A(k+1, 2) & ... & A(k+1, n(k+1)) ==> P
```

The methodology for learning the Bayes net is as follows:

1. The set of implications I(k) is transformed into a data table D.
2. This data table is used to generate a Bayes net, using standard Bayes net structure and parameter learning methods

Once the Bayes net has been learned, then the standard Bayes net belief propagation method can be used to perform the needed inference.

The weakness of this approach is that Bayes net structure learning is a slow process that must be re-done whenever the relevant set S of implications changes considerably. However, once the Bayes net is in place, then the inference step may be done very rapidly.

The only nonstandard step here is the production of the data table D from the set S of implications. This is to be done as follows:

- The first column header of D corresponds to the target predicate P. The subsequent column headers of D correspond to the set of unique predicates found on the left-hand side of the predicates I(k) in S.
- The truth-value strengths s_k are discretized into a small number d of values, say 11 values. So, each s_k is replaced with a value sd_k, which is of the form $r/(d\text{-}1)$ for $r = 0,\ldots,(d\text{-}1)$.
- Then, each implication I(k) is used to generate d rows of the data table D.
  - In $sd_k * d$ of these rows the first (P) column has a 1; in the remaining $(1\text{-}\ sd_k)*d$ rows it has a 0.
  - In the other columns a 0 value is placed except for those columns corresponding to the predicates that appear as A(k, i) on the left-hand side of I(k).

for instance, we have 200 implications in S, involving 30 predicates, and a discretization of strength values into 10 values, then we have a binary data table with 31 columns and 2000 rows. This data table may then be fed into a standard Bayes net learning algorithm.

Finally, what do we do if we have an implication of the form

```
(A(k, 1) || A(k, 2))& ... & A(k, n(k)) ==> P <s_k>
```

with a disjunction in it? In general, we could have any Boolean combination on the left-hand side.

If there's just a single disjunction as in the above example, it can be tractably handled by merely doubling the number of data rows created for the implication, relative to what would be done if there were no disjunction. But more broadly, if we have a general Boolean formula, we need to take a different sort of approach. It seems we can unproblematically create rows for the data table by, in the creation of each row, making a random choice for each disjunction. It might be better (but is not initially necessary) to put the formulas in a nice normal form such as Holman's ENF (Holman 1990; Looks 2006) before doing this.

9.5 Trail-Free Inference via Function Optimization or Experience-Driven Self-Organization

(Co-authored with Cassio Pennachin)

Finally, this section suggests a yet more radical approach to the problem of doing inference on a large number of logical relationships, with minimal error yet reasonably rapidly. The approach suggested is not trivial to implement, and has not yet been experimented with except in very simple cases, but we are convinced of its long-term potential. We will focus here on the simple experiments we've carried out, then at the end talk a little about generalization of the ideas.

In the experiments we conducted, we dealt only with Inheritance relationships joining ordinary, first-order terms. Also, we considered for simplicity Atoms labeled with simple (strength, weight of evidence) = (s, w) truth-values. The only inference rules we considered were deduction, revision, and inversion. For revision we used an iterated revision rule called "revision-collapse," which acts on an Atom and collapses the multiple versions of the Atom into a single Atom (using the revision TV formula).

Define a selection operator as a function that maps a list of Atoms into a proper subset of that list; e.g., "select all relationships," "select all relationships that have multiple versions." Define an inference protocol as a function that maps a list of Atoms into another one, defined as a combination of selection operators and inference rules. An inference protocol is a special case of an inference control strategy

– the specialization consisting in the fact that in a protocol there is no adaptiv wherein the conclusion of one inference affects the choice of what inferences are done next. A real-world inference engine needs a sophisticated inference control strategy – a topic that will be raised in Chapter 13. But for the simple experiments described here, a basic and rigid inference protocol was sufficient. There is a "global revision" inference protocol that is defined as: Select all Atoms with multiple versions, and apply revision-collapse to them.

The standard (forward chaining) FOI inference protocol without node probability inference (SP) is defined as follows:

- Select all InhRelationships, and apply inversion to them.
- Do global revision.
- Select all InhRelationship pairs of the form (A --> B, B --> C), and apply deduction to them.
- Do global revision.

The standard protocol with node probability inference (SPN) is defined by adding the two steps at the end:

- Do node probability inference.
- Do global revision.

One may generate a vector of TVs corresponding to a set of Atoms. Using (s, w) truth-values, this becomes an n-dimensional vector of pairs of floats.

Taking the protocol SP as an example, given an Atom-set AS(0) we may define

```
AS(n+1) = SP( AS(n) )
```

as the Atom-set resulting from applying the protocol SP to the Atom-set AS(n). This series of Atom-sets implies a series of TV-vectors

```
V(n) = v(AS(n))
```

where v is the mapping taking Atom-sets into TV-vectors.

Potentially the series of AS's may converge, yielding a fixed point

```
AS* = SP(AS*)
```

In this case we may say that the corresponding TV-vector V* represents a TV assignment that is *consistent according to the inference protocol SP* (or, SP-consistent). A problem, however, is that V* need not be the closest SP-consistent TV-vector to the original TV vector V(0).

Similarly, we may define an e-consistent (e-SP-consistent) Atom-set as one for which

$$\left|v(AS) - v^*(SP(AS))\right| < e$$

The degree of SP-inconsistency of AS may be defined as the minimum e for which the above inequality holds. If AS's degree of SP-inconsistency is 0, then AS is SP-consistent.

If the series AS(*n*) converges to a fixed point, then for any e there is some *N* so that for *n*>*N*, AS(*n*) is e-consistent. However, these e-consistent Atom-sets along this series need not be the closest ones to AS(0).

What is the most sensible way to compute the distance between two TV-vectors (and, by extension, between two Atom-sets)? One approach is as follows. Assume we have

$$V1 = ((s11, c11), (s12, c12), \ldots, (s1n, c1n))$$
$$V2 = ((s21, c21), (s22, c22), \ldots, (s2n, c2n))$$

Then we may define

$$d(V1, V2) = \sum_i \left[c1i * c2i * |s1i - s2i|\right]$$

This is a 1-norm of the difference between the strength components of V1 and V2, but with each term weighted by the confidences of the components being differenced. One approach to inference, then, is to explicitly search for an Atom-set whose SP-consistent (or e-SP-consistent) TV-vector is maximally close to AS(0).

This might in a sense seem less natural than doing iterated inference as in the series AS(*n*) described above. However, the iterated inference series involves a compounding of errors (the errors being innate in the independence assumptions of the deduction and revision TV formulas), which can lead to a final attractor quite distant from the initial condition in some cases. One way to avoid this is to introduce inference trails – a strategy which effectively stops the iteration after a finite number of steps by using a self-halting inference protocol that incorporates trails, and is hence more complex than SP or SPN described above. Another approach is the one suggested here: use an optimization algorithm to find a conclusion TV-vector satisfying specified criteria.

There are several possible ways to formalize the optimization problem. Define *x* to be a semi-minimum of a function F if there is some neighborhood *N* about *x* so that for all *y* in *N*, $F(y) \geq F(x)$; or an e-semi-minimum if $F(y) \geq F(x) - e$. Then we may, for instance, look for:

- The AS that is a semi-minimum of SP-inconsistency that is closest to AS(0)

- For fixed e, the AS that is an e-semi-minimum of SP-inconsistency t closest to AS(0)
- The AS that minimizes a weighted combination of inconsistency and distance from AS(0)
- The AS that minimizes a weighted combination of {the smallest e for which AS is an e-semi-minimum of SP-inconsistency} and distance from AS(0)

Let us call this approach to the dynamics of inference optimization-based inference (OBI).

One obvious weakness of the OBI approach is its computational cost. However, the severity of this problem really depends on the nature of the minimization algorithm used. One possibility is to use a highly scalable optimization approach such as stochastic local search, and apply it to a fairly large Atom-set as a whole. This may be viable if the Atom-set in question is relatively static.

However, another interesting question is what happens if one takes a large Atom-set and applies OBI to various subsets of it. These subsets may overlap, with the results of different OBI runs on different overlapping subsets mediated via revision. This approach may be relatively computationally tractable because the cost of carrying out OBI on small Atom-sets may not be so great.

For instance, one experiment we carried out using this set-up involved 1000 random "inference triangles" involving 3 relationships, where the nodes were defined to correspond to random subsets of a fixed finite set (so that inheritance probabilities were defined simply in terms of set intersection). Given the specific definition of the random subsets, the mean strength of each of the three inheritance relationships across all the experiments was about .3. The Euclidean distance between the 3-vector of the final (fixed point) relationship strengths and the 3-vector of the initial relationship strengths was roughly .075. So the deviation from the true probabilities caused by iterated inference was not very large. Qualitatively similar results were obtained with larger networks. Furthermore, if one couples together a number of inference triangles as described in the above paragraph, revising together the results that different triangles imply for each shared relationship, then one obtains similar results but with lower correlations – but still correlations significantly above chance.

9.5.1 Approximating OBI via Experience-Driven Self-Organization

The same thing that OBI accomplishes using an optimization algorithm may be achieved in a more "organic" way via iterated trail-free inference in an experiential system coupled to an environment. Harking back to the "experiential semantics" ideas of Chapter 3, one way to think about OBI is as follows. Considering the ase studies above, of simple terms interrelationshiped by Inheritance relationships, we may imagine a set of such terms and relationships dynamically updated according to the following idea:

1. Each term is assumed to denote a certain perceptual category.
2. For simplicity, we assume an environment in which the probability distribution of co-occurrences between items in the different categories is stationary over the time period of the inference under study.
3. We assume the collection of terms and relationships has its probabilistic strengths updated periodically, according to some "inference" process.
4. We assume that the results of the inference process in Step 3 and the results of incorporating new data from the environment (Step 2) are merged together ongoingly via a weighted-averaging belief-revision process.

This kind of process will have qualitatively the same outcome as OBI. It will result in the network drifting into an "attractor basin" that is roughly probabilistically consistent, and is also pretty close to the data coming in from the environment.

The key thing in this picture is the revision in Step 4: It is assumed that, as iterated inference proceeds, information about the true probabilities is continually merged into the results of inference. If not for this, Step 3 on its own, repeatedly iterated, would lead to noise amplification and increasingly meaningless results. But in a realistic inference context, one would never simply repeat Step 3 on its own. Rather, one would carry out inference on a term or relationship only when there was new information about that term or relationship (directly leading to a strength update), or when some new information about other terms/relationships indirectly led to inference about that term-relationship. With enough new information coming in, an inference system has no time to carry out repeated, useless cycles of inference on the same terms/relationships – there are always more interesting things to assign resources to. And the ongoing mixing-in of new information about the true strengths with the results of iterated inference prevents the pathologies of circular inference, without the need for a trail mechanism.

The conclusion pointed to by this line of thinking is that if one uses an inference control mechanism that avoids the repeated conduction of inference steps in the absence of infusion of new data, issues with circular inference are not necessarily going to be severe, and trails may not be necessary to achieve reasonable term and relationship strengths via iterated inference. Potentially, circular inference can occur without great harm so long as one only does it when relevant new data is coming in, or when there is evidence that it is generating information. This is not to say that trail mechanisms are useless in computational systems – they provide an interesting and sometimes important additional layer of protection against circular inference pathologies. But in an inference system that is integrated with an appropriate control mechanism they are not necessarily required. The errors induced by circular inference, in practice, may be smaller than many other errors involved in realistic inference.

Chapter 10: Higher-Order Extensional Inference

Abstract In this chapter we significantly extend the scope of PLN by explaining how it applies to what we call "higher-order inference" (HOI) – by which we mean, essentially, inference using variables and/or using relationships among relationships. One aspect of HOI is inference using quantifiers, which is deferred to the following chapter because in the PLN approach it involves its own distinct set of ideas.

10.1 Introduction

The term "higher-order" is somewhat overloaded in mathematics and computer science; the sense in which we use it here is similar to how the term is used in NARS and in functional programming, but different from how it is traditionally used in predicate logic. In the theory of functional programming, a "higher order function" refers to a mathematical function whose arguments are themselves functions. Our use of the term here is in this spirit. PLN-HOI involves a diverse assemblage of phenomena, but the essential thing that distinguishes HOI from FOI is a focus on relationships between relationships rather than relationships between simple terms. As an example of relationships between relationships, much of PLN-HOI deals with relationships between logical predicates, where the latter may be most simply understood as functions that take variables as arguments (though there is also an alternative interpretation of predicates using combinatory logic, in which predicates are higher-order functions and there are no variables. Here we will generally adopt the variable-based approach, though the concepts presented also apply in the combinatory-based approach). Logical relationships between predicates (variable-bearing or no) are then an example of relationships among relationships.

In the context of PLN-HOI, the boundary between term logic and predicate logic becomes somewhat blurry and the approach taken is best described as an amalgam. Logical predicates are used centrally and extensively, which is obviously reminiscent of predicate logic. On the other hand, the basic term logic inference rules (such as Aristotelian deduction) also play a central role and are extended to handle Implication relationships between predicates rather than Inheritance relationships between simple terms. This synthetic approach has been arrived at primarily for practical reasons; sticking with pure predicate or term logic (or pure combinatory logic, or any other available "pure" approach) appeared to introduce unnatural complexity into some intuitively simple inferences.

present approach, inferences that appear intuitively simple tend to come out formally simple (which is one of the qualitative design guidelines underlying PLN as a whole).

PLN-HOI, like PLN-FOI but more so, is a set of special cases. There are many HOI inference rules. However, there are not as many new truth-value formulas as in FOI, though there are some. The PLN approach is to reduce higher-order inferences to first-order inferences wherever possible, via mapping predicates into their SatisfyingSets. This allows the theory of probabilistic first-order inference to naturally play the role of a theory of probabilistic higher-order inference as well.

We will begin here by giving some basic PLN rules for dealing with relationships between predicates and between first-order relationships. Then as the chapter develops we will broach more advanced topics, such as Boolean operators, variable instantiation, and combinatory logic based higher-order functions. The following chapter continues the story by extending HOI to handle universal, existential, and fuzzy quantifiers.

10.2 PLN-HOI Objects

Before getting started with higher-order inference proper, we briefly review some of the knowledge representation that will be used in the HOI rules. Most of this material was presented earlier in the chapter on knowledge representation and is briefly repeated here with a slightly different slant.

A Predicate, in PLN, is a special kind of term that embodies a function mapping Atoms into truth-values. The truth-value of a predicate is similar to the term probability of an ordinary "elementary" term. It can be estimated using term probability inference, or evaluated directly. When evaluated directly it is calculated by averaging the truth-value of the predicate over all arguments. In practice, this average may be estimated by using all known Evaluation relationships in the reasoning system's memory regarding this predicate. In cases where lots of effort is available and there aren't enough Evaluation relationships, then a random sampling of possible inputs can be done.

All the inference rules for predicates work the same way for multi-argument predicates as for single-argument predicates. This is because a multi-argument predicate may be treated as a single-argument predicate whose argument is a list (multi-argument predicates may also be handled via currying, but discussion of this will be deferred till a later subsection). Let, for instance, (Ben, Ken) denote the list alternatively denoted

```
List Ben Ken
```

Then (using the inference rule for combining Evaluation and Implication relationships, to be given below) we may say things like

```
Evaluation kiss (Ben, Ken)
Implication kiss marry
|-
Evaluation marry (Ben, Ken)
```

an inference that goes just the same as if the argument of kiss were an elementary term rather than a list.

We will also make use of relationships involving variables along with, or instead of, constant terms, for instance

```
Evaluation do (Ben, $X)
```

Such relationships may be implicitly considered as predicates, and are treated symmetrically with other predicates in all PLN rules. The above example may be treated as a predicate with a single variable, represented by the notation $X. This is a predicate that evaluates, for each argument $X, the degree to which it's true that Ben does X.

Next, as well as predicates we will also make some use of objects called Schemata, which are terms that embody functions mapping Atoms into things other than TruthValue objects. Schemata come along with their own relationship types, Execution and ExecutionOutput (ExOut for short), which have the semantics that if a Schema F embodies a function *f* so that

$$f(x) = y$$

then

```
ExOut f x = y
Execution f x y
```

In short, ExOut denotes function application, whereas Execution records the output of a function application.

We have introduced a few additional relationship types to make HOI simpler – some of these mentioned above. There is

```
SatisfyingSet
```

which relates a predicate to the set of elements that satisfy it; e.g.,

```
SatisfyingSet isAnUglyPerson UglyPeople
```

Then there's

```
Quantifier
```

with subclasses

```
ForAll

ThereExists
```

e.g., to express

```
"There is some woman whom every man loves"
```

one can write

```
ThereExists $X:
 AND
    Inheritance $X woman
    ForAll $Y:
          Implication
                Inheritance $Y man
                Evaluation loves ($Y,$X)
```

Finally, there's

```
BooleanRelationship
```

with subclasses

```
AND, OR and NOT
```

which also has some other more specialized subclasses to be introduced later. We have just used AND, for instance, in the above example. The most straightforward use of the Boolean relationships is to join relationships or predicates, but there are also special variants of the Boolean relationships that exist specifically to join terms.

10.2.1 An NL-like Notation for PLN Relationships

To make the text output of our current PLN implementation easier to understand, we have created a simple pseudo-NL syntax for describing PLN relationships. While not syntactically correct English in most cases, we have found this syntax generally much easier to read than formal notation. This notation will be used in some of the following examples. The following table illustrates the NL-like syntax (which we will use in examples for the remainder of this section):

Formal representation	NL-like representation
Inheritance *Concept mushroom* *Concept fungus*	*Mushroom is fungus*
Implication *Predicate isMushroom* *Predicate isFungus*	*If isMushroom then isFungus*
ListRelationship *Concept yeast* *Concept mushroom* *Concept mold*	*(yeast, mushroom, mold)*
Evaluation *Predicate friendOf* *List* *Concept kenji* *Concept lucio*	*FriendOf(kenji,lucio)*
And *Concept yeast* *Concept mushroom* *Concept mold*	*And(yeast, mushroom, mold)*
Not *Concept yeast*	*Not yeast*
ExOut *Schema friendOf* *Concept lucio*	*The result of applying friendOf to lucio*

10.3 Where HOI parallels FOI

The simplest kind of HOI is the kind that exactly mirrors FOI, the only difference being a substitution of predicates for simple terms and a substitution of higher-order relationship types for their corresponding first-order relationship types. An example of this kind of inference is the following deduction involving the predicates is_Brazilian and eats_rump_roast:

```
Implication is_Brazilian is_ugly
Implication is_ugly eats_rump_roast
|-
Implication is_Brazilan eats_rump_roast
```

Similar things may be done with Equivalence relationships. This kind of HOI follows the FOI rules, where Implication behaves like Inheritance, Equivalence behaves like Similarity, and so forth. Further examples will be omitted here due to their obviousness, but many examples of such inference will occur in the sample inference trajectories given in Chapter 14 when we discuss application of the current PLN implementation to making a virtual agent learn to play fetch,.

10.4 Statements as Arguments

Next we discuss a case that lies at the border between first-order and higher-order inference. Some commonsense relationships, such as "know" and "say," take statements as arguments. Some commonsense concepts, such as "fact" and "sentence," take statements as instances. Because the embedded statements in these examples are treated just like ordinary terms, the PLN FOI rules treat them as other terms, though from a semantic point of view the inference is higher-order.

For example, "Ben believes that the Earth is flat" can be represented as

```
Evaluation
    believe
    ( Ben ,  (Inheritance Earth flat_thing))
```

"The Earth is flat is a ridiculous idea" then becomes

```
Inheritance <tv1>
    Inheritance Earth flat_thing
    ridiculous_idea
```

By induction, from this statement and Ben's belief in the flatness of the Earth, one may conclude

```
Inheritance <tv2>
    SatisfyingSet (believe_curry Ben)
    ridiculous_idea
```

and thus

```
Evaluation <tv3>
    believe_curry Ben
    ridiculous_idea
```

which becomes

```
believe Ben ridiculous_idea <tv3>
```

i.e., "Ben believes in ridiculous ideas."

10.5 Combining Evaluations and Implications

Combining an Evaluation and an Implication yields an Evaluation, as in the example

```
Evaluation is_American Ben <tv1>
Implication is_American is_idiot <tv2>
|-
Evaluation is_idiot Ben <tv3>
```

The truth-value strength here is obtained via deduction, for reasons to be elucidated below. For simple truth values, the weight of evidence is a discounted version of the value obtained via deduction, where the discount (which we'll call a "multiplier") is the product of the M2ICountMultiplier and the I2MCountMultiplier. In the case of indefinite truth values, we replace the count multiplier with I2MIntervalWidthMultiplier and M2IIntervalWidthMultiplier. These width multipliers keep the means of the truth values constant while simply increasing the interval widths to account for losses in confidence.

This is a "heuristic" form of inference whose correctness is not guaranteed by probability theory, due to the use of the I2M and M2I heuristic inference rules. These rules are approximatively valid for "cohesive concepts" whose members share a lot of properties; they're badly invalid for random concepts.

We now run through the detailed logic of this inference rule, using the above illustrative example. First, by the definition of SatisfyingSet, we have:

```
Execution SatisfyingSet is_American  Americans <1>
|-
Member Ben Americans <tv1>
```

(where "Americans" is the set of all Americans). We may also define the SatisfyingSet of is_idiot:

```
Execution SatisfyingSet is_idiot idiots <1>
```

By the M2I (MemberToInheritance) heuristic we have

```
Member Ben Americans <tv1>
|-
Inheritance {Ben} Americans <tv2>
```

where tv2 has the same strength as tv1 but, depending on truth-value type, a discounted count or wider interval. By deduction we then have

```
Inheritance {Ben} Americans <tv2>
Inheritance Americans idiots <tv3>
|-
Inheritance {Ben} idiots <tv4>
```

The I2M heuristic then yields

```
Inheritance {Ben} idiots <tv4>
|-
Member Ben idiots <tv5>
```

where tv5 has the same strength as tv4 but a discounted count (i.e. a wider interval).

From the definition of SatisfyingSet we then have

```
Evaluation is_idiot Ben <tv5>
```

10.6 Inference on Boolean Combinations

Extensional Boolean operators may be defined on predicates in a straightforward way. For instance, if P and Q are Boolean predicates, then the degree to which (P AND Q) is true of a set S may be defined as the percentage of members x of S for which both P(x) and Q(x) are true. If P and Q are non-crisp predicates, then the degree to which (P AND Q)(x) is true may be defined in terms of fuzzy intersection, and the degree to which (P AND Q)(S) is true may be defined truth-value of the fuzzy intersection (P AND Q) evaluated over the set S.

Extending these notions to terms is also straightforward, because a Concept term A may be considered in terms of a predicate whose SatisfyingSet is A. In this way, Boolean operators on terms may be defined by means of Boolean operators on predicates. This is equivalent to the more direct approach of defining Boolean operation on Concept terms as set operations – according to which, e.g., the intersection of two Concept terms A and B is the Concept whose member-set is the intersection of the member-sets of A and B.

10.6.1 Inference on Boolean Compounds of Predicates

First we give some heuristic rules for dealing with Boolean operators. These rules are generally bad approximations, so they are to be applied only when time is short or when there is no data to do direct evaluation. We illustrate them with simple examples:

Inference Rule	Truth-value Strength Formula

`is_Ugly`
`AND is_Male is_Ugly`
`|-`
`isMale`
	`isMale.tv.s = (isMale & isUgly).tv.s / isUgly.tv.s`

`is_Ugly`
`OR is_Male is_Ugly`
`|-`
`is_Male`
	`isMale.tv.s = (MU-U)/(1-U)`
	`MU = (isMale OR isUgly).tv.s`
	`U = isUgly.tv.s`

`isMale`
`isUgly`
`|-`
`isMale AND isUgly`
	`(isMale & isUgly).tv.s = isMale.tv.s * isUgly.tv.s`

`isMale`
`is Ugly`
`|-`
`isMale OR isUgly`
	`(isMale & isUgly).tv.s = isMale.tv.s + isUgly.tv.s - isMale.tv.s * isUgly.tv.s`

`isMale`
`|-`
`NOT isMale`
	`(NOT isMale).tv.s = 1-isMale.tv.s`

e subtle point about these rules is that a complex Boolean expression can be evaluated by applying these simple rules in many possible orders. Each order of evaluation allows the usage of different knowledge in memory. For instance, if we have

```
is_Male AND is_Ugly AND is_Stupid
```

and the memory contains information about

```
is_Male AND is_Ugly
```

and

```
is_Ugly AND is_Stupid
```

then the inference controller must choose whether to do the evaluation as

```
(is_Male AND is_Ugly) AND is_Stupid
```

or as

```
is_Male AND (is_Ugly AND is_Stupid)
```

Or, the system can do the evaluation both ways and then use revision to merge the results! Optimizing the evaluation order for large expressions is a hard problem. The approach taken in the current implementation is to use all evaluation orders for which the memory contains reasonably confident data, and then revise the results.

10.6.2 Boolean Logic Transformations

Next, there are logical transformation rules, which are given in ordinary Boolean logic notation as:

$$(P \vee Q) \Leftrightarrow (\neg P \Rightarrow Q)$$
$$(P \wedge Q) \Leftrightarrow \neg(P \Rightarrow \neg Q)$$
$$(P \Leftrightarrow Q) \Leftrightarrow ((P \Rightarrow Q) \wedge (Q \Rightarrow P))$$

and which may be executed in PLN without change of truth-value.

For example,

```
isMale OR isUgly <s,w>
```

may be transformed into

```
Implication <s,w>
    ExOut NOT isMale
    isUgly
```

or else

```
Implication <s,w>
    P
    isUgly
```

where P is the predicate whose internal function is “NOT isMale”; or else, with variables,

```
Implication <s,w>

    Evaluation (ExOut NOT isMale) X
    Evaluation isUgly X
```

10.6.2.1 Boolean Operators for Combining Terms

As already noted, the simple Boolean operators discussed above, though presented in the context of combining predicates, may also be used to combine elementary terms, according to the strategy of considering a term as the SatisfyingSet of a predicate and then combining the terms via combining these predicates. Or, equivalently, one may introduce separate operators for acting on terms, defined, e.g., by

extensional intersection: (A AND_{Ext} B)

```
Equivalence
    Member x (A AND_Ext B)
    (Member x A) AND (Member x B)
```

extensional difference: (A $\text{MINUS}_{\text{Ext}}$B)

```
Equivalence
    Member x (A MINUS_Ext B)
    (Member x A) AND NOT (Member x B)
```

extensional union: (A OR_{Ext}B)

```
Equivalence
    Member x (A OR_Ext B)
    (Member x A) OR (Member x B)
```

A number of conclusions immediately follow from these definitions; e.g.,

```
Equivalence
    Subset x (A AND_Ext B)
    (Subset x A) AND (Subset x B)

Implication
    (Subset x A) OR (Subset x B)
    Subset (A AND_Ext B) x
```

To understand the importance of the latter implication, consider a simple example. Assume there is a term for "Bank" and a term for "American Company," and we want to make a new term that includes the common instances of the two (so intuitively what we get is "American Bank"). Obviously, the extension of the new term should be built by the *intersection* of the extensions of "Bank" and "American Company" as arguments, because a term is an instance of "American Bank" if and only if it is a "Bank" and it is an "American Company." This is what the extensional intersection does.

On the other hand, as the common subset of "Bank" and "American Company," the supersets of "American Bank" include both the supersets of "Bank" and those of "American Company," so the intension of the new term should be the *union* of the intensions of "Bank" and "American Company." For example, "American Bank" is a kind of "Financial Institution" (because "Bank" is a kind of "Financial Institution"), though "American Company" is not a kind of "Financial Institution." If we use the intersection (or union) operator for both the extension and intension we can still get a term, but it has no natural interpretation in terms of its components. Fortunately, elementary set theory behaves in the appropriate way and we have the implication given above, which says that the extensional intersection belongs to categories defined as the disjunction of the sets of categories to which its components belong.

Regarding extensional difference, note that

```
Equivalence
    Subset x (A MINUS_AsymExt B)
    (Subset x A) AND NOT (Subset x B)

Equivalence
    Subset (A MINUS_AsymExt B) x
    Subset A x
```

The asymmetric extensional difference *A* $MINUS_{Ext}$ *B* contains things that are in the extension of A but not in the extension of B. Whether these things will belong to the same categories as B or not, we can't say, which is why there is no *Subset B x* term in the second implication given above.

Regarding union, we have:

```
Equivalence
    Subset x (A OR_AsymExt B)
    (Subset x A) OR (Subset x B)

Equivalence
    (Subset A x) OR (Subset B x)
    Subset (A OR_AsymExt B) x
```

These term logical operators may be used inside an inference system in two different ways. They may be used as operators inside larger implications, or they may be used as heuristics for "seeding" the creation of new concepts. For instance, given the existence of the concepts of American Company and Bank, implications of the form

```
Implication
    American_Company AND_AsymMix Bank
    Y
```

may be learned. Or, a new concept C may be formed, initially defined as

```
American_Company AND_AsymMix Bank
```

but given the possibility of "drifting" over time via forming new relationships, gaining new members, etc. Both strategies have their value.

10.7 Variable Instantiation

As we've noted, variables are not necessary to PLN, in the sense that PLN-HOI also works with combinatory logic operators, which avoid the need for variables via use of higher-order functions. We will discuss this approach in Section 10.9 and elsewhere below. However, in our practical implementation of PLN we have found variables quite useful, and so we have used them in several of the previous subsections of this chapter on HOI, and will continue to use them in later sections and chapters; and we will now describe some rules specifically oriented toward manipulating variables.

riable instantiation refers to the inference step wherein one takes a relationship involving variables and then inserts specific values for one of the variables. Of course, instantiating multiple variable values in a relationship can be handled by repeated single-variable instantiations. A useful truth-value function for variable instantiation can be derived as a consequence of several truth-value functions previously derived.

In general, suppose we have

```
F($X) <tv>
```

where $X is a variable, and we instantiate this to

```
F(A) <tv1>
```

The most direct way to do this is to observe that

```
s1 = P( F($X) | $X = A)
```

The most accurate and conceptually coherent method for estimating this conditional probability is to include both direct and indirect evidence. We can accomplish this by interpreting

```
s1 = P( F($X) | Similarity $X A)
```

so that we have

```
Implication
    AND
          Evaluation F $X
          Similarity A $X
    Evaluation F A
```

To turn this interpretation into a concrete formula, it suffices to break down the proposed inference step into a series of substeps that are equivalent to previously discussed inference rules. What we need is to find a truth-value for the conclusion of the inference:

```
Similarity A B
Evaluation F B
|-
Evaluation A B
```

But this inference can be transformed, via two applications of the M2I inference rule, into

```
Similarity A B
Inheritance B (SatisfyingSet F)
|-
Inheritance A (SatisfyingSet F)
```

which can be transformed into the following via the sim2inh rule:

```
Inheritance A B
Inheritance B (SatisfyingSet F)
|-
Inheritance A (SatisfyingSet F)
```

And now we have arrived at an ordinary deductive inference.

So according to this line of reasoning, the truth-value of the variable-instantiation inference conclusion should be:

```
F(A).tv = I2M ( deduction( M2I(F.tv),
                sim2inh( (Similarity A $X).tv) )
```

Note that in the above derivation, the assumption is made that the SatisfyingSet of F is a "cohesive" set; i.e., that the elements within it tend to be fairly similar to each other. Without this assumption this approach doesn't work well, and there's no way to say anything about F(A) with a reasonably high confidence. PLN automatically accounts for this lack of confidence via the count or interval width multipliers in the M2I/I2M rule, which depend on how cohesive the set involved is. More cohesive sets involve smaller multipliers. The right multiplier for each set may be computed empirically if elements of the set are known, or else it may be computed by inference via the assumption that similar sets will tend to have similar cohesions and therefore similar interval width multipliers.

As a particular example of the above inference rule, consider the following two inferences, expressed in pseudo-NL notation:

1)

```
If isDead($X) then there exists $Y so that
    killed($Y,$X) <[0.1, 0.3], 0.9, 10>
|-
If isDead(Osama) then there exists $Y so that
    killed($Y,Osama) tv1
```

2)

```
If isDead($X) & isTerrorist($X), then there exists $Y
    so that killed($Y,$X) <[0.5, 0.7] , 0.9, 10>
```

```
If isDead(Osama) & isTerrorist(Osama), then there
    exists $Y so that killed($Y,Osama) tv2
```

These inferences include quantifiers, which won't be discussed until the following chapter, but that's not the main point we need to consider here. Assuming

```
[ Similarity $X Osama ].tv = <[0,15, 0.25], 0.95, 10>
```

and assuming M2I/I2M interval width multipliers of 1.5 for both inferences, then we obtain

```
tv1 = <[0.123024, 0.486104], 0.9, 10>
tv2 = <[0.235028, 0.570565], 0.9, 10>
```

On the other hand, if we use the fact that the premise of the second inference involves a more cohesive set than the premise of the first inference, then that means that the second inference involves a lesser M2I/I2M interval width multiplier than the first one. In fact the set of $X who are dead terrorists is much smaller and much more cohesive than the set of $X who are dead entities. For dead terrorists (inference 2) we may plausibly assume an M2I/I2M interval width multiplier of 1.2. In this case we obtain the revised value

```
tv2 = <[0.259443, 0.574509], 0.9, 10>
```

10.7.1 Implications Between Non-Variable-Bearing Relationships

Now we present an interesting, somewhat conceptually subtle application of the variable-instantiation formula. We show that it is possible to construct higher-order relationships such as implications that relate simple relationships without variables – and without any higher-order functions being involved. The semantics of these relationships, however, is subtle and involves "possible worlds." The truth-values of these relationships can be derived via a combination of the possible-worlds interpretation with the variable-instantiation truth-value formula given above. For simplicity, we will discuss this formula in the context of simple truth-values, but extrapolation to indefinite truth-values is not difficult.

As an example, consider

```
Implication
    Ben is ugly
    Ben eats beef
```

How can we define the truth-value of this?

First, we suggest, the most useful way to interpret the expression is:

```
Implication
    In possible universe C, Ben is ugly
    In possible universe C, Ben eats beef
```

Note that this reinterpretation has a variable – the possible universe C – and hence is interpretable as a standard Implication relationship. But we may interpret, e.g.,

```
In possible universe C, Ben is ugly
```

as meaning

```
Inheritance (Ben AND "possible universe C ") ugly
```

In other words, one may view Ben as a multiple-universe-spanning entity, and look at the intersections of Ben with the possible universes in the multiverse.

Suppose then that we know

```
Implication
    Inheritance $X ugly
    Inheritance $X beef_eater
```

Variable instantiation lets us derive from this

```
Implication
    Inheritance (Ben AND "possible universe C ") ugly
    Inheritance (Ben AND "possible universe C")
        beef_eater
```

which is equivalent by definition to

```
Implication
    Inheritance Ben ugly
    Inheritance Ben beef_eater
```

So, in short, we may quantify inheritance between non-variable-bearing relationships using the variable instantiation formula.

10.8 Execution Output Relationships

As noted above, if we have a function that outputs an Atom, then we use an ExecutionOutput (ExOut for short) relationship to refer to the Atom that's output,

Execution relationship to denote the ternary relationship between schema, input, and output. Here we describe a few basic rules for manipulating ExOut and Execution relationships.

First, note that we have the tautology

```
Execution $S $X (ExOut $S $X) <1,1>
```

For example, if we want to say "The agent of a killing situation is usually Armenian" then we may say

```
Implication
    Inheritance $x killing
    Inheritance
          ExOut Agent $x
          Armenian
```

which is equivalent to

```
Implication
    Inheritance $x killing
    Implication
          Execution Agent $x $y
          Inheritance $y Armenian
```

In this example, the ExOut relationship allows us to dispense with an additional relationship and an additional variable. There is a general rule

```
Equivalence <1,1>
    Evaluation F (ExOut A B)
    AND
          Execution A B $Y <1,1>
          Evaluation F $Y
```

that may be used to convert expressions with ExOuts into expressions without them.

Finally, an alternative way of saying

```
ExOut Agent $X
```

is to construct a predicate P whose internal predicate-function evaluates "Agent $X". Then we have

```
Implication
    Inheritance $X killing
```

```
Inheritance
      Evaluation P $X
      Armenian
```

10.9 Reasoning on Higher-Order Functions

PLN-HOI is capable of dealing with functions or relationships of arbitrarily high order, with relationships between relationships ... between relationships, or functions of functions ... of functions. This property implies that a fully capable PLN can be created without any use of variables at all, due to the possibility of using combinatory logic to represent mathematical expressions that would normally be denoted using variables, in an entirely variable-free way. In practice, it seems that the use of variables is often worthwhile because it allows more compact representation of useful higher-order relationships than would be possible without them. However, there are also many cases where curried relationships of the style common in combinatory logic and combinatory logic based programming languages are useful within PLN, due to their own ability to represent certain types of relationship with particular compactness.

As a simple example of a case where the curried representation is semantically natural, consider the schema "very" which maps a predicate P into the predicate "very P." Then we might have

```
Implication
    ExOut very $X
    $X
```

from which we can draw the conclusion e.g.

```
Implication
    ExOut very isMale
    isMale
```

Note that the original statement is equivalent to

```
Implication
    ExOut (ExOut very $X) $Y
    ExOut $X $Y
```

Alternately, we can do the same thing with combinators rather than variables. To make this work nicely we want to use the curried version of Implication, in which, e.g.,

```
(Evaluation (ExOut Implication_curried A) B).tv
```

where the standard Implication relationship might also be called Implication_list, since it is in effect a predicate that takes a list argument. The above example becomes

```
Evaluation
    ExOut
          Implication_curried
          ExOut very $X
    ExOut I $X
```

using the I combinator. To eliminate the variable we use the definitions

```
S f g x = (fx) (gx), so
   B f g x = f(gx)
```

which lets us reason as follows, using Imp as a shorthand for Implication_curried:

```
(Imp (very x)) (I x)  =
((B Imp very) x) (I x) =
S (B imp very) I x
```

In term and relationship notation, this final result would be written

```
ExOut
      ExOut
             S
             A
      I

A :=
ExOut
      ExOut
             B
             imp
      very
```

Note the cost of removing the variables; we've introduced a few extra terms and relationships. Whether the variable-free or variable-bearing form is better depends on the kind of reasoning you're doing. Removing variables allows one to use simpler inference control strategies, but generally increases the size of the structures being reasoned on. The variable-free, curried approach is useful only in cases where this size increase is not too severe.

10.10 Unification in PLN[1]

The concept of unification is an important part of theoretical computer science, and also plays a key role in AI; for example, via its central role in the Prolog language (Deransart et al 1996) and its role in computational linguistics (in unification grammars) (Roark and Sproat 2007.) This section explores the manifestation of unification in PLN, which is relatively simple but different from usual treatments of unification due to the lack of focus on variables and the central role played by uncertainty.

To illustrate the nature of unification in PLN, let us introduce an example involving a Predicate P relating the expression values of two genes (XYZ and ABC) at different times:

```
P($T) :=
Implication
    expression(XYZ,$T)>.5 AND expression(ABC,$T+1)<.5
    expression(XYZ,$T)<.2
```

and a related Predicate Q:

```
Q :=
Implication
     expression(XYZ, $T)>.9
     expression(XYZ, $T)<.4
```

Now, it is not hard to compute the truth-value of the relationship

```
Implication P Q
```

by referring to the actual underlying data points, assuming one has access to this data. But what if one wants to compute this truth-value without doing this, just by manipulating P and Q? To do this, one has to "unify" the two expressions.

We may write the two expressions as

```
Implication L1 R1
Implication L2 R2
```

It is possible to find L and R, with nontrivial truth-values, so that

```
Implication L L1 <t1>
Implication L L2 <t2>
```

[1] This section was coauthored with Guilherme Lamacie.

```
                     1 <t3>
Implication R R2 <t4>
```

also with nontrivial truth-values. Specifically,

```
L($T) := [ expression(XYZ, $T) > .9 ]
R($T) := [ expression(XYZ, $T) < .2 ]
```

will work. Induction then yields

```
Implication L1  L2 <t5>
Implication R1 R2 <t6>
```

which then yields

```
Implication <t7>
    Implication L1  R1
    Implication L2  R2
```

by HOI rules.

The “unification” step here is the determination of L and R; these terms “unify” the two original expressions. In this case, because of the specialized nature of the expressions, the unifying expressions are not hard to find. In general, finding unifying expressions is not so simple.

10.10.1 PLN Unification versus Prolog Unification

It may not be immediately clear how this process we’ve called “unification” relates to unification as conventionally done in mathematical logic, for instance in the Prolog programming language. A few brief comments in this regard may be valuable. We will use Prolog notation freely, and so this section may not be fully comprehensible to the reader who’s not familiar with Prolog; for necessary background the reader is referred to (Spivey 1996.)

A typical Prolog unification might involve a query such as

```
? son(A, sam), sister(B, becky), grandpa (A, B)
```

and a database consisting of

```
father(ted, ben)
father(victor, ted)
father(sam, victor)
    male(X) ← father(X, Y)
```

```
grandpa(X, Z) ← father(X, Y),  father(Y, Z)
sister(ben, becky)
son(X, Y) ← father(Y, X), male(X)
```

To process this, *son(A, sam)* is unified with *son(X, Y)*, with the substitution *$s1 = {A/X_1, Y/sam}*, yielding a third expression *son(X_1,sam)*. This expression is then replaced with *father(sam, X_1), male(X_1)*. The first Predicate *father(sam, X_1)* is unified with *father(sam, victor)*, by substitution *$s2 = {X_1/victor}*. Given this substitution, the next step is to try to solve *male(victor)*, which is done by unifying it with *male(X)*, by *$s3 = {X/victor}* and then solving the tail Predicate *father(victor, Y)* (i.e., unifying it with *father(victor, ted)*). Now, the next Predicate in the query *sister(B, becky)* is unified with *sister(ben, becky)*, by *$s4 = {B/ben}*. Note that substitutions *$s1*,*$s2* and *$s4* cause the last Predicate in the query to be *grandpa(victor, ben)*. This can be now solved by unification with *grandpa(X, Z)*, by *$s5 = {X/victor, Z/ben}*, and the third expression produces *grandpa(victor, ben)*, which leads to the tail Predicates *father(victor, Y)*, *father(Y, ben)*, which are further solved analogously, thus yielding a proof of *grandpa(Victor,Ben)*.

In this type of unification, one compares two expressions to each other and tries to find variable values that will work in both expressions. The trick is in the order of processing of expressions; for this, Prolog uses a simple depth-first search with backtracking.

PLN unification is a bit different, because there are not necessarily any variables, and implication relations are probabilistic rather than crisp. However, the above example can easily be done in PLN in several ways. Here we will show how it can be done by reformulating each expression in a variable-free way. For instance

```
grandpa(X, Z) ← father(X, Y),  father(Y, Z)
```

becomes

```
Implication (B father father) grandpa
```

(since B father father x y = father (father x) y). We then have a "database" consisting of

```
Evaluation (father Ted) Ben
Evaluation (father Victor) Ted
Evaluation (father Sam) Victor
Implication (B father father) grandpa
Evaluation (sister Ben) Becky
Implication son (C father)
```

What is the "query"? Well, seeing that

```
son(a, Sam) =(son a) Sam = C I Sam (son a) =
                                  B (C I Sam) son a

sister(b, Becky) = B (C I Becky) sister b
```

the query is then just a search for terms a, b so that

```
a ∈ SatisfyingSet (B (C I Sam) son )
b ∈ SatisfyingSet (B (C I Becky) sister)
b ∈ SatisfyingSet ( grandpa a)
```

To put this more elegantly, we can introduce an argument list (a,b) and then define

```
listify grandpa (a, b) = grandpa a b
```

We are then looking for an element of

```
SatisfyingSet( listify grandpa ) ∩
SatisfyingSet( (B (C I Sam) son, B (C I   Becky) sister) )
```

i.e., for an element that makes the Predicate

```
AND
    listify grandpa
    ( (B (C I Sam) son, B (C I Becky) sister)
```

output the value True.

Instead of doing classical variable unification, what happens to find an element of the SatisfyingSet of this Predicate function? First, the "sister" term in the term *B (C I Becky) sister* is matched with the "sister" term in the known fact

```
    Evaluation (sister Ben) Becky
```

So *Ben* is tried out as an argument of *B (C I Becky) sister* and is indeed found to cause this term to evaluate to True. Then, "son" is found not to match to anything concrete, but is found in the formula

```
Implication son (C father)
```

This formula is then used to create information such as

```
Evaluation (son Ben) Ted
Evaluation (son Ted) Victor
```

```
Evaluation (son Victor) Sam
```

The last of these then matches the term B (C I Sam) son.

On the other hand, there is no concrete match for *grandpa* in the given knowledge base, but there is a rule *grandpa* = *B father father*, which matches directly with the knowledge

```
Evaluation (father Ted) Ben
Evaluation (father Victor) Ted
Evaluation (father Sam) Victor
```

In this way the system is eventually led to discover that the pair

```
(Ben, Victor)
```

lies in the appropriate SatisfyingSet.

This example illustrates the general principle that, in a variable-free context, logical unification comes down to finding elements in SatisfyingSets that are defined by intersections of the SatisfyingSets corresponding to Predicates of interest.

The chief difference between this example and the earlier example involving gene expression relationships is absence of probabilistic truth-value in the present example. In Prolog-like unification, two terms either match or they don't – i.e., an item is either a member of a given SatisfyingSet or it's not. In general PLN unification, this is not the case; there can of course be degrees of membership, which makes the unification process yet more computationally difficult.

10.10.2 General Unification

The general problem of unification, in a PLN context, is as follows. Given two Predicates R and S, one wants to find a Predicate T so that

```
SatisfyingSet(T)=SatisfyingSet(R)∩SatisfyingSet(S)
```

and *SatisfyingSet(T)* is as large as possible.

Of course, one can nominally solve the problem by simply forming the Predicate

```
T = R AND S
```

But of course in doing this one has no real idea of the truth-value of R AND S, unless one makes a possibly fallacious independence assumption. In some cases, after all, R and S may contradict each other, so that SatisfyingSet(T) is empty. The problem of unification, then, is basically the problem of making a dependence-estimate of the truth-value of the compound *R AND S*. In practice, rather than evaluating *T = R AND S* explicitly, one may seek T so that

```
Implication T (R AND S)  <t>
```

with $t<1$, but with the property that the truth-value of T can be relatively readily assessed based on the system's knowledge.

There is no known *generally effective* algorithm for solving this problem. There are only more or less plausible heuristics. For instance, suppose that R and S are both conjunctive compounds of the form

$$R = R_1 \text{ AND } \ldots \text{ AND } R_n$$
$$S = S_1 \text{ AND } \ldots \text{ AND } S_n$$

Then one can seek to unify each R_i with some S_j. Of course, this is not all that easy. First of all there are n^2 pairings to look at. And for each one of these pairings one is stuck with a smaller, but possibly still difficult, unification problem.

Here the terms R_i and S_j are effectively serving as the analogue of "variables" in Prolog-style unification. There are two profound difficulties here that are not seen in Prolog, however. First, there is no normal form for logical expressions here, so we don't know which R_i should match with which S_j. Second, rather than just crisply substituting one "variable" for another, we have to evaluate probabilistic implications (truth-values of relationships of the form *Implication T* (R_i, *AND* S_j).

The gene expression example given above is actually an easier case than the conjunctive compound case. A brief re-analysis of this example may be instructive at this point. There we had two examples of the form

$$\text{Implication } L_1 \; R_1$$
$$\text{Implication } L_2 \; R_2$$

Because these were obviously of the same format, it was reasonable to seek to unify L_1 and L_2, and R_1 and R_2 respectively. No consideration of a large number of permutations was required. Because of the particular form of the terms in this case, the intersections were not hard to find. For instance

```
R1 AND R2 :=
expression(XYZ,t)<.2  AND  expression(XYZ,t)<.4
```

This is an expression that can be simplified easily since

$$\text{Implication } R_1 \; R_2$$

is given, and it's known that

```
AND( A, B)
Implication A B
|-
A
```

so that

```
R1 AND R2 = expression(XYZ,t)<.2
```

The case

```
L1 AND L2
:=
AND
    expression(XYZ, t)>.5 AND expression(ABC, t+1)<.5
    expression(XYZ, t)>.9
```

is not so simple, however. It can immediately be simplified to

```
L1 AND L2
:=
expression(XYZ, t)>.9  AND expression(ABC, t+1)<.5
```

To go further from here one has to either make some wild guesses or revert to actually evaluating this predicate on real data. The same goes for the further, more speculative reduction

```
Implication
    expression(XYZ, t) > .9
    L1 AND L2
```

In practice, then, we propose that unification problems can be approached in PLN via a combination of explicit quantitative predicate evaluation, and transformation using the earlier-given predicate transformation rules. The general rule is: keep on transforming until it doesn't work anymore, and then start evaluating – where "doesn't work anymore" means, essentially, that a precipitous drop in confidence has occurred as a result of risky transformations.

10.11 Inference on Embedded Relationships

Two crucial cases of higher-order knowledge, requiring special treatment, are *hypothetical* and *contextual* knowledge. These are different cases, but closely related. Each of them represents knowledge that is stored in an AI system's memut implicitly understood not to pertain to the universe at large in a straightforward way. PLN can handle these kinds of knowledge easily, so long as the proper knowledge representation structures are put in place.

10.11.1 Hypothetical Knowledge

As a simple, archetypal example of hypothetical knowledge, consider the statement "Matt believes the Earth is flat." Before presenting the way this is dealt with in PLN, we will discuss a couple of *incorrect* ways of presenting it to explain why we have chosen the course we have. First, the statement could facilely be represented as

```
Evaluation believe Matt (Inheritance Earth flat)
```

The problem with this, however, is that it involves entering a bogus piece of knowledge

```
Inheritance Earth flat
```

into the system. One might try to get around this problem by judiciously inserting truth-values, such as

```
Evaluation <1>
    believe Matt (Inheritance Earth flat <0>)
```

But this is not a good approach; immediately one faces difficulties. This expression has two possible interpretations:

```
"Matt is certain that the Earth is not flat"
"Matt is certain that the Earth is flat, but No-
   vamente believes it is not flat"
```

One might deal with this particular case by just accepting the second interpretation. But this isn't really a general way out, because it doesn't give one a way to say "Matt is pretty certain that the Earth is almost surely flat." If one tries to render this latter statement as

```
Evaluation <.7>
    believe Matt (Inheritance Earth flat <.9>)
```

then one is inserting the bogus knowledge *(Inheritance Earth flat <.9>)* into the system and there is no easy workaround.

The solution we have taken in PLN is to explicitly introduce a representat "hypotheticalness" into the picture. We do this with the Hypothetical Atom, a unary relationship which has the informal semantics that

```
Hypothetical A
```

means "don't assume A is true, just assume 'Hypothetical A' is true."

The interesting thing is, it's not necessary to give PLN any kind of refined, explicit semantics for dealing with Hypothetical (sometimes Hyp for short). Rather, Hyp is a marker. Inferences such as

```
Evaluation believe Matt (Hyp (Inheritance Earth flat))
Evaluation believe Matt (Hyp (Inheritance Mars Earth))
|-
Evaluation believe Matt (Hyp (Inheritance Mars flat))
```

can be made automatically, just as if Hyp were any other relationship type. The special rule required is, quite simply, a rule stating that before drawing a conclusion

```
L1
L2
|-
L3
```

it should be determined whether either L1 or L2 is hypothetical (is pointed to by a Hyp) or not. If so, then the inference is not done – though inferences of the form

```
P (Hyp L1)
P (Hyp L2)
|-
P (Hyp L3)
```

may be done, for appropriate P, as in the preceding example. This "inference inhibition" rule is the full explicitly encoded semantics of Hypothetical.

An interesting question is whether, for example, the entry of the relation

```
Matt believes ( Hyp (Inheritance unicorns cute))
```

into the system causes the strength of the truth-value of the *unicorn* Term to be increased. In other words, if the system finds it's often valuable to posit hypothetical unicorns, should this increase the system's estimate of the probability of unicorns in "its world"? One may introduce a system parameter determining the amount hypothetical mention of an Atom increases the TruthValue of that Atom, but ultimately this sort of thing must be dealt with differently in different contexts, hence it requires the learning of appropriate cognitive schemata.

The combination of hypothetical and nonhypothetical information may also be useful, although it must be handled with care. For instance, suppose we want to reason

```
Matt believes the Earth is flat
Flat things are not round
|-
Matt believes the Earth is not round
```

Properly enough, there is no way to do this directly. Rather, we need an additional premise such as

```
Implication A (Matt believes Hyp(A))
```

or

```
Implication
    (A AND (Inheritance A C))
    (Matt believes Hyp(A))

Implication (Inheritance flat NOT(round)) C
```

where C is some category of knowledge in which the system is fairly sure Matt believes, in spite of his eccentric beliefs in other domains.

Human beings are quite adept at forming such categories C, and thus at managing the overlapping but nonidentical belief systems of themselves and other human beings. This is a skill that seems to be learned in middle childhood, along with other aspects of advanced inference and advanced social understanding.

10.11.2 Higher-Order Statements and Judgments

The topic of hypothetical knowledge brings to the fore another issue that has more general relevance: the difference between *higher-order statements* and *higher-order judgments*. A higher-order statement is a relationship that treats its component relationships as truth-value-free entities. A higher-order judgment is a relationship that treats its component relationships as truth-valued entities.

An example of higher-order judgment is:

```
Evaluation believe Matt (Hyp (Inheritance Earth flat
<.9>))
```

This statement is saying that Matt believes the Earth is flat with a .9 strength.

Similarly, to say that Cassio does not believe the Earth is flat we could say, for instance,

```
Evaluation
    believe
    List
          Cassio
          Hypothetical
                Inheritance Earth flat <.01>
```

The presumption here is that the TruthValue of *(Inheritance Earth flat)* is part of the entity being fed to the *believe* Term as an argument.

On the other hand, what would be an example of a higher-order statement? When we say

```
Implication
    AND
          Inheritance Earth flat
          Inheritance Earth planet
    Inheritance planet flat
```

we actually don't mean

```
Implication
    AND
          InheritanceEarth flat <t1>
          Inheritance Earth planet <t2>
    Inheritance planet flat <t3>
```

In other words, we don't mean

```
The earth is flat with TruthValue t1
The earth is a planet with TruthValue t2
|-
Planets are flat with TruthValue t3
```

where t_1, t_2, t_3 are the actual truth-values of the respective statements in the system. Inference *could* construct statements like this, but then the truth-value t_3 produced by inference from t_1 and t_2 might not be the actual truth-value of *Inheritance planet flat* in the system. In fact it generally will not be, unless this inference step is the only thing that has ever affected the truth-value of this relationship.

In our above examples of hypothetical relationships, we were assuming "higher-order judgment" (HOJ) style inference – that is, we were assuming rela-ips were being referred to with truth-values intact. On the other hand, in our reasoning about flat planets, we were assuming "higher-order statement" (HOS) style inference; that is, we were assuming relationships were being referred to without specific truth-values attached.

It seems that in hypothetical inference we often do want to assume the HOJ case. Otherwise, the HOS case is the most common case but not the exclusive case.

There are, however, some cases where one might want to make non-hypothetical higher-order judgments. For instance, to represent "I am positive that countries that are rich in oil are moderately wealthy," one might use

```
Implication <.99>
    AND
          Inheritance $X country
          Inheritance $X oil_rich
    Inheritance $X wealthy <.6>
```

Here the <.99> is for the "positive"; the <.6> is for the "moderately." Notationally, one may distinguish the two cases implicitly as follows: if a truth-value marker <t> is included in denoting an Atom that is part of another relationship, one may assume that HOJ is implied.

Implementationally, to make the distinction one needs to introduce some kind of marker to distinguish the two cases. The marker should attach to logical relationships; e.g., Implication in the above example. We have chosen to make "higher-order statement" the default case, and require a special marker to denote "higher-order judgment." Where extra notational explicitness is required we may denote this as, for instance,

```
Implication_HOJ <.99>
    AND
          Inheritance X country
          Inheritance X oil_rich
    Inheritance X wealthy <.6>
```

We have been focusing on cases involving relationships because this is the most common kind of higher-order judgment, but something similar can happen with Terms; e.g., one can say

```
Inheritance_HOJ
    ugly <.1>
    beautiful <.8>
```

meaning that things that are not very ugly are often very beautiful. However, this example is better said (to within a decent degree of approximation) as

```
Inheritance
    ugly
    NOT beautiful
```

Generally speaking, in the situations we have analyzed, cases where inter-term HOJ relationships are the most convenient alternative are even rarer than similar cases with interrelationship HOJ relationships.

10.11.3 Contextual Knowledge

Hypothetical knowledge is knowledge that the reasoning system may not believe at all; contextual knowledge, on the other hand, is knowledge that the system believes only in certain circumstances. For instance, suppose we want to say "Ben is competent in the domain of mathematics; Ben is incompetent in the domain of juggling." What we mean here is something like

```
Implication
    Ben AND doing_mathematics
    competent

Implication
    Ben AND doing_juggling
    NOT competent
```

Here we are saying that instances of Ben who are doing mathematics are competent (at what they are doing), whereas instances of Ben who are doing juggling are not competent (at what they are doing).

Of course, this is only one among many possible representations of the posited conceptual relationships in terms of terms and relationships. In reality, there may be no single term for "competent" or "doing_juggling", the given HOS's may be represented as collections of HOJ's, etc. We have just chosen a particularly simple possible representation for expository purposes.

The case of contextual knowledge is so common that we believe it is worthwhile to have a special representational mechanism just for dealing with it. For this reason we introduce Context, with the semantics

```
Context C (Hyp (R X Y))
:=
R (X AND_Ext C) (Y AND_Ext C)

Context C (Hyp (R X Y Z))
:=
R (X AND_Ext C) (Y AND_Ext C) (Z AND_Ext C)
```

So for example

```
Context
     doing_mathematics
     Hyp (Inheritance Ben competent)

Context
     doing_juggling
     Hyp
             NOT (Inheritance Ben competent)
```

Using the above definition of Context these translate into

```
Implication
     Ben AND_Ext doing_mathematics
     competent AND_Ext doing_mathematics

Implication
     Ben AND_Ext doing_juggling
     NOT
             competent AND_Ext doing_juggling
```

which are equivalent to the forms given previously.

The value of the Context "macro" is that it simplifies inferences such as

```
Context
     doing_mathematics
     Hyp (Inheritance Ben competent)
Context
     doing_mathematics
     Hyp (Inheritance Gui competent)
|-
Context
     doing_mathematics
     Hyp (Similarity Ben Gui)
```

This kind of inference could be done without Context, using the explicit AND_{Ext}-heavy representations of the premise relationships. However, the existence of Context has the effect of pushing the system to make inferences of this type, whereas otherwise such inferences would be carried out only occasionally as part of the generic process of inference on Predicates. Now, an advanced AI system without Context could be expected to learn from experience that inference on Predicates of the "Context-ish" variety is useful, and create cognitive schema biasing the inference process toward combining Predicates of this type. However, we feel that explicitly pushing the system to combine Contexts inferentially wil big help in getting any PLN-based reasoning system to the point where it's able to do this kind of advanced cognitive schema creation.

10.12 An Application of PLN-HOI to Inference Based on Natural Language Information Extraction

In this section we summarize an example of PLN-HOI that was published in (Ikle and Goertzel, 2007), which was previously carried out using PLN inference rules together with alternate inference formulas acting directly on SimpleTruth-Values. The analysis given in (Ikle and Goertzel, 2007) was a modification of an earlier treatment given in (Goertzel et al 2006), which utilized heuristic weight of evidence formulas; the newer treatment utilized indefinite probabilities. What we present here is a small and partial "inference trail" that is part of a larger trail described in (Goertzel et al 2006), which was part of an experiment in integrative NLP using an integrated system that

- Parsed the sentences in PubMed abstracts using a grammar parser
- Transformed the output of the grammar parser into logical relationships using a collection of expert rules
- Performed probabilistic logical inference on these logical relationships using PLN

The system, called Bioliterate, was created under a contract from the NIH Clinical Center and was specifically tuned to infer relationships between genes, proteins and malignancies. The overall inference of which this example is a part is depicted qualitatively in the following table, which shows the premises in the form of the actual sentences present in PubMed automatically extracted and used as premises.

Premise 1	Importantly, bone loss was almost completely prevented by p38 MAPK inhibition.
Premise 2	Thus, our results identify DLC as a novel inhibitor of the p38 pathway and provide a molecular mechanism by which cAMP suppresses p38 activation and promotes apoptosis.
Conclusion	DLC prevents bone loss. cAMP prevents bone loss.

The premises depicted in this table were automatically converted into sets of l relationships using a set of hand-coded expert rules embodying linguistic knowledge, and PLN was used to draw the conclusion via forward chaining inference. The following table depicts a fragment of the overall inference trail, which basically gathers together a number of mutually relevant relationships within a single conjunction.

Rule	Premises	Conclusion
Abduction	Inh inhib1, inhib <[.95, 1], .9> Inh inhib2, inhib <[.95, 1], .9>	Inh inhib1, inhib2 <[0, 0.221708], .9>
Similarity Substitution	Eval subj (prev1, inhib1) <[.95, 1], .9> Inh inhib1, inhib2	Eval subj (prev1, inhib2) <[.05, .44], .9>
Deduction	Inh inhib2, inhib Inh inhib, causal_event <[.95, 1], .9>	Inh inhib2, causal_event <[.856, .998], .9>
AND	Inh inhib2, causal_event Inh prev1, causal_event <[.854, .998], .9> Eval subj (prev1, inhib2) Eval subj (inhib2, DLC) <[.95, 1], .9>	AND <[0.0625, 0.358972], 0.9> Inh inhib2, causal_event Inh prev2, causal_event Eval subj <prev2, inhib2> Eval subj <inhib$_2$, DLC>

In the overall inference trail this conjunction is then used as a premise to a subsequent inference step, which uses unification to conclude that Eval subj (prev1, inhib2); i.e., that the prevention mentioned in Premise 1 is the subject of the inhibition mentioned in Premise 2. We next use reverse truth-value conversion to convert the full <[*L*, *U*], *b*, *k*> truth-values into <*s*, *n*, *b*> triples for all of our premises, as well as for our conclusion. The truth-value conversion results are shown in the table:

Premise	<[*L*, *U*],*b*>	<*s*,*n*,*b*>
Inh inhib1 inhib	<[0.95, 1], 0.9>	<.988, 28, .9>
$inhib_1$	<[0.95, 1], 0.9>	<.988, 28, .9>
inhib	<[0, 0.168], 0.8>	<.083, 8, .8>
$inhib_2$	<[0, 0.032], 0.8>	<.016, 15, .8>
Conclusion		
AND	<[0.06, 0.4], .9>	<.214, 6, .9>

The heuristic approach reported previously produced a truth-value of <1, 0.07> for the final conjunction. The indefinite probabilities approach produced a truth-value of <[0.0625, 0.358972], 0.9)> , or <.2143, 6> in <s,n> form, and <.2143, .358> in <s,w> form. These two results are not logically contradictory at all, but the indefinite probabilities result is more informative. The prior heuristic formulas told us that there is very little evidence supporting the contention that the strength of the conclusion is near 1. The indefinite probabilities formulas tell us (more usefully) that there is a reasonable though not overwhelming amount of evidence that the strength of the conclusion is near .21. Being more principled, the indefinite probabilities method is guaranteed, if the underlying distributional assumptions are realistic, to focus on the most interesting part of the conclusion truth-value distribution.

Chapter 11: Handling Crisp and Fuzzy Quantifiers with Indefinite Truth-Values

Abstract In this chapter, we exemplify the use of indefinite probabilities in the handling of inference involving crisp universal and existential quantifiers and also fuzzy quantifiers. The treatment is novel, involving third-order probabilities, but appears to give intuitively sensible results in specific examples.

11.1 Quantifiers in Indefinite Probabilities

We have discussed a variety of rules for inference on predicates, but haven't yet broached the subtlest aspect, which is inference on quantified expressions. This is a case where the indefinite probabilities approach bears considerable fruit, allowing us to articulate a conceptually clear (though complex) approach that seems to cut through the confusion we perceive to exist in much of the literature on uncertain inference with quantifiers.

The approach we outline is a subtle one. The best way we have found to handle quantifiers within the indefinite probabilities framework is to introduce another level of complexity and utilize third-order probabilities. To understand this, we first consider the problem of "direct evaluation" of the indefinite truth-values of universally and existentially quantified expressions.

Building on the ideas from the previous chapter, we solve this problem via a semantic approach that is considerably conceptually different from the one standardly taken in formal logic. Normally, in logic, expressions with unbound variables are not assigned truth-values; truth-value assignment comes only with quantification. In our approach, however, as exemplified in the previous chapter, we assign truth-values to expressions with unbound variables, yet without in doing so binding the variables. This is unusual but not contradictory in any way; an expression with unbound variables, as a mathematical entity, may certainly be mapped into a truth-value without introducing any mathematical or conceptual inconsistency. This allows one to define the truth-value of a quantified expression as a mathematical transform of the truth-value of the corresponding expression with unbound variables, a notion that is key to our approach.

This unusual semantic approach adds a minor twist to the notion that our approach to uncertain inference on quantified expressions reduces to standard crisp inference on quantified expressions as a special case. The twist is that our approach reduces to the standard crisp approach in terms of truth-value assignation for all expressions for which the standard crisp approach assigns a truth-value.

ver, our approach also assigns truth-values to some expressions (formulas with unbound variables) to which the standard crisp approach assigns no truth-value.

Following up on this semantic approach, we will now explain how, if we have an indefinite probability for an expression *F(t)* with unbound variable t, summarizing an envelope E of probability distributions corresponding to *F(t)*, we may derive from this an indefinite probability for the expression "ForAll *x*, *F*(*x*)." (Having carried out the transform in this direction, it will then be straightforwardly possible to carry out a corresponding transform in reverse.) The approach we take here is to consider the envelope E to be part of a higher-level envelope E1, which is an envelope of envelopes. The question is, then, given that we have observed E, what is the chance (according to E1) that the true envelope describing the world actually is almost entirely supported within [1-e, 1], where the latter interval is interpreted to constitute "essentially 1" (i.e., e is the margin of error accepted in assessing ForAll-ness), and the phrase "almost entirely supported" is defined in terms of a threshold parameter?

Similarly, in the case of existential quantification, we want to know the indefinite probability corresponding to "ThereExists *x*, *F*(x)." The question is, then, given that we have observed E, what is the chance (according to E1) that the true envelope describing the world actually is *not* entirely supported within [0, e], where the latter interval is interpreted to constitute "essentially zero" (i.e., e is the margin of error accepted in assessing ThereExists-ness)?

The point conceptually is that quantified statements require you to go one level higher than ordinary statements. So if ordinary statements get second-order probabilities, quantified statements must get third-order probabilities. And the same line of reasoning that holds for "crisp" universal and existential quantifiers turns out to hold for fuzzy quantifiers as well. In fact, in the approach presented here, crisp quantifiers are innately considered as an extreme case of fuzzy quantifiers, so that handling fuzzy quantifiers doesn't really require anything extra, just some parameter-tuning.

The following sections elaborate the above points more rigorously.

11.2 Direct Evaluation of Universally and Existentially Quantified Expressions

We first consider the case of the direct evaluation of universally quantified expressions, an inference rule for which the idea is as follows: Given an indefinite truth-value for *F*(*t*), we want to get an indefinite TV for *G* = ForAll *x*, *F*(*x*).

The roles of the three levels of distributions are roughly as follows. The first- and second-order levels play the role, with some modifications, of standard indefinite probabilities. The third-order distribution then plays the role of "perturbing" the second-order distribution. The idea is that the second-order distribution represents the mean for the statement $F(x)$. The third-order distribution then various values for x, and the first-order distribution gives the sub-distributions for each of the second-order distributions.

The process proceeds as follows:

1. Calculate [lf1,uf1] Interval for the third-order distribution. This step proceeds as usual for indefinite probabilities: see (Iklé and Goertzel 2008; Iklé et al 2007). Given L, U, k, and b, set $s = 0.5$. We want to find a value for the variable *diff* so that the probability density function defined by $$f(x) = \frac{(x-L1)^{ks}(U1-x)^{k(1-s)}}{\int_{L1}^{U1}(x-L1)^{ks}(U1-x)^{k(1-s)}dx}$$ where $L1=L-diff$ and $U1=U+diff$ is such that $$\frac{\int_{L1}^{L}(x-L1)^{ks}(U1-x)^{k(1-s)}dx}{\int_{L1}^{U1}(x-L1)^{ks}(U1-x)^{k(1-s)}dx} = \frac{1-b}{2}$$ and $$\frac{\int_{U}^{U1}(x-L1)^{ks}(U1-x)^{k(1-s)}dx}{\int_{L1}^{U1}(x-L1)^{ks}(U1-x)^{k(1-s)}dx} = \frac{1-b}{2}.$$ Once one of these last two integrals is satisfied, they both should be. Alternatively, one can find *diff* for which $$\frac{\int_{L}^{U}(x-L1)^{ks}(U1-x)^{k(1-s)}dx}{\int_{L1}^{U1}(x-L1)^{ks}(U1-x)^{k(1-s)}dx} = b.$$
2. At present we are using only beta distributions for the desired "third-order" distribution family. To generate vectors of means for perturbed $F(x)$ values, we first generate a vector of length $n1$ of random values chosen from a standard beta distribution. Next we scale the random means to the interval [lf1,uf1] using a linear transformation.
3. Now we use the same procedure as in Step 1 to generate symmetric intervals $[\mathrm{lf2}[\mathrm{i}],\mathrm{uf2}[\mathrm{i}]]$ for each of the means found in Step 2. These intervals are now the desired [L1, U1] intervals for the third-order distributions.
4. For each mean for the third-order distributions, we generate a sub-distribution. These sub-distributions represent the second-order distributions.
5. We next generate first-order distributions with means chosen from the second-order distributions.

w we determine the percentage of elements in each first-order distribution that lie within the interval [1-e, 1]. Recall that we are using the interval [1-e, 1] as a "proxy" for the probability 1. The goal here is to determine the fraction of the first-order distributions that are almost entirely contained in the interval [1-e, 1]. By "almost entirely contained" we mean that the fraction contained is at least `proxy_confidence_level` (PCL).

7. Finally, we find the conclusion <[*L*,*U*],*b*> interval. For each of the third-order means, we calculate the average of all of the second-order distributions that are almost entirely contained in [1-e, 1], giving a list of $n1$ elements, *probs*, of probabilities. We finally find the elements of *probs* corresponding to quantiles using $L = \left\lfloor \frac{n1(1-b)+1}{2} \right\rfloor$ and $U = \left\lfloor \frac{n1(.5+b)+1}{2} \right\rfloor$

Given the above, it is simple to obtain the ThereExists rule through the equivalence

ThereExists x, $F(x)$ $\Leftrightarrow \neg$[ForAll x, $\neg F(x)$] .

11.3 Propagating Indefinite Probabilities through Quantifier-Based Inference Rules

As well as "directly evaluating" quantifiers in the manner of the previous section, it is also necessary within a logical reasoning system to carry out various quantifier manipulations. We now discuss a variety of transformation rules that work on quantifiers, drawn from standard predicate logic.

First, we have already seen that what is called "the rule of existential generalization" holds in the indefinite probabilities framework (this is just a reformulation of what we have called "direct evaluation" of existentially quantified expressions, above):

1) *F*(*c*) <[*L*,*U*], *b*, *k*>
|-
∃ \$*x*, *F*(\$*x*) <[*L*,*U*], *b*, *k*>
where *c* may be any expression not involving \$*x*.

Next, consider universal specification:

2) ∀ \$*x*, *F*(\$*x*) <[*L*,*U*], *b*, *k*>
|-
F(*c*) <[*L*,*U*, *b*, *k*>
where *c* is any expression not involving \$*x*.

To see that universal specification also holds with indefinite probabilities, given the truth-value above for ForAll \$*x*, *F*(\$*x*), we can obtain an indefinite truth-value for $F(t)$. We then use the mean of $F(t)$ over all values t, as a heuristic appro tion to $F(c)$ for a given value c.

We have already also seen, at least implicitly, that all the standard quantifier exchange formulas hold for indefinite probabilities:

3) $\neg(\exists x)F(x) \Leftrightarrow (\forall x)\neg F(x)$
$(\exists x)\neg F(x) \Leftrightarrow \neg(\forall x)F(x)$
$\neg(\exists x)\neg F(x) \Leftrightarrow (\forall x)F(x)$
$(\exists x)\neg F(x) \Leftrightarrow \neg(\forall x)\neg F(x)$

For our last transformation rule, we consider the operation of removing constants from within existential quantifiers. In predicate logic we have that:

4) $\forall x$: G AND $F(x) = G$ AND $\forall$ x: $F(x)$

Unlike the case for crisp predicate logic however, this rule is not, in general, true using indefinite probabilities. For example, consider the following set of premises with parameter settings e=0.5 and PCL=0.7: truth-value for G = <[0.45, 0.46], 0.9, 10> and truth-value for $F(x)$ = <[0.71, 0.72], 0.9, 10>. Then the result for $\forall x$: G AND $F(x)$ becomes <[0.0, 0.04913], 0.9, 10>, while that for G AND $\forall x$: $F(x)$ is <[0.23046, 0.28926], 0.9, 10>. On the other hand, we note that a different set of premises can yield similar results from the two approaches. Assuming the same parameter values for e and PCL, and truth-values for both $F(x)$ and G of <[0.99, 1.0], 0.9, 10> gives a result of <[0.98331, 0.99626], 0.9, 10> using $\forall x$: G AND $F(x)$, and a similar result of <[0.98344, 0.99620], 0.9, 10> using G AND $\forall x$: $F(x)$.

For insight into what is happening here, we view $H(F)(t) = G$ AND $F(t)$ as a distortion of the distribution of F. In addition, if $J(F) = \forall x$ J(x), then $J(F)$ is a nonlinear distortion of F, so that even though $H(F)$ is a linear distortion, it need not commute with J. An obvious and interesting question is then: Under what combination of premise values and parameter settings do the operators H and J "almost" commute? Due to space considerations we defer a thorough study of that question to a future paper. It does appear, however, that premise values near 1 lead to better commutativity than do values farther from 1.

11.4 Fuzzy Quantifiers

Analyzing the indefinite probabilities approach to the quantifiers *ForAll* and *ThereExists*, it should be readily apparent that indefinite probabilities provide a natural method for "fuzzy" quantifiers such as *AlmostAll* and *AFew*.

In our discussion of the *ForAll* rule above, for example, the interval [PCL, 1] represents the fraction of bottom-level distributions completely contained in the interval [1-e, 1]. Recall that the interval [1-e,1] represents a proxy for probability 1.

analogy with the interval [PCL, 1] representing the *ForAll* rule, we can introduce the parameters **lower_proxy_confidence** (LPC) and **upper_proxy_confidence** (UPC) so that the interval [LPC, UPC] represents an *AlmostAll* rule or an *AFew* rule. More explicitly, by setting [LPC, UPC] = [0.9, 0.99], the interval could now naturally represent *AlmostAll*.

Similarly, the same interval could represent *AFew* by setting LPC to a value such as 0.05 and UPC to, say, 0.1.

Through simple adjustments of these two proxy confidence parameters, we can thus introduce a sliding scale for all sorts of fuzzy quantifiers. Moreover, each of these fuzzy quantifiers is now firmly grounded in probability theory through the indefinite probabilities formalism.

11.5 Examples

To further elucidate the above formalism, we now consider two examples. For our first example, we consider an example drawn from [15], which is there called the "crooked lottery" and extensively discussed:

$$\left[\neg \text{ThereExists } x \middle| \text{ Winner}(x) \rightarrow \text{false}\right] \&$$

$$\left[\text{ThereExists } y \middle| \left[\text{ForAll } x \middle| \left(\text{Winner}(x) \parallel \text{Winner}(y)\right) \rightarrow \text{Winner}(y)\right]\right]$$

The first clause is intended to represent the idea that everyone has a nonzero chance to win the lottery; the second clause is intended to represent the idea that there is one guy, y, who has a higher chance of winning than everybody else. In [15] Halpern examines various formalisms for quantifying uncertainty in a logical reasoning context and assesses which ones can provide a consistent and sensible truth-value evaluation for this expression.

To evaluate the truth-value of this expression using indefinite probabilities, suppose we assume the truth-value for Winner(x) is <[0.05, 0.1], 0.9, 10>. For the second clause we also assume that the truth-value for Winner(y) is <[0.25, 0.5], 0.9, 10> and that the truth-value for the implication $\left(\text{Winner}(x) \parallel \text{Winner}(y) \rightarrow \text{Winner}(y)\right)$ is <[0.8, 0.9], 0.9, 10>. With these assumptions, we then vary the values of the parameters e and PCL for the ThereExists rule to generate the following graphs of the resulting truth-value intervals. Note that the parameter values e and PCL used in the ForAll rule were the complements, 1-e and 1-PCL, of the values used in ThereExists.

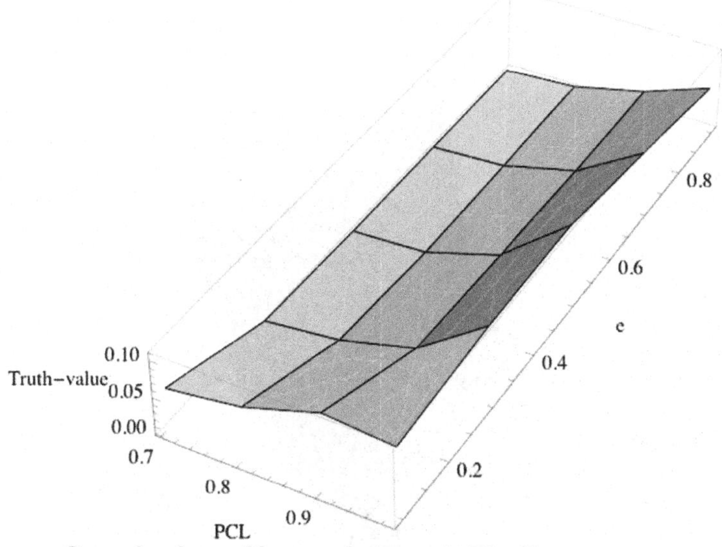

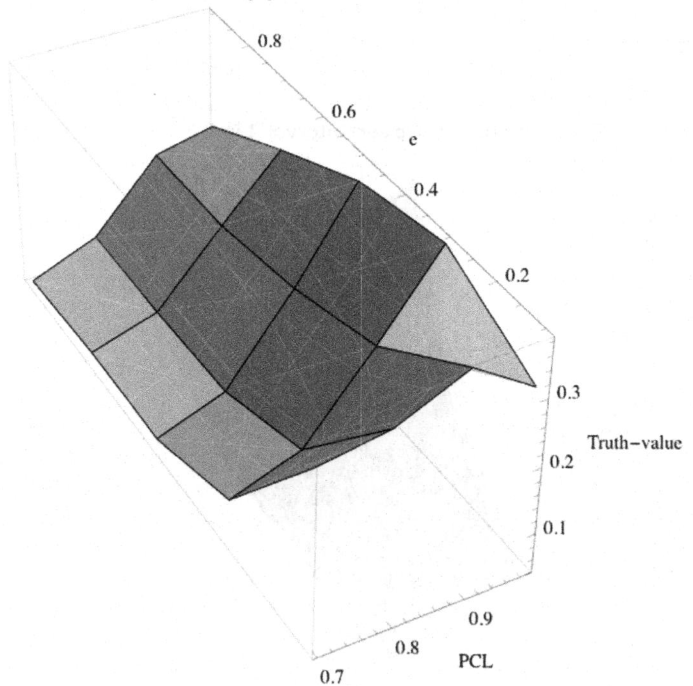

The intermediate results for each of the two main clauses provide insight into the interaction between these two clauses. First the results for clause 1: $\left[\neg \text{ThereExists } x \middle| \text{ Winner}(x) \rightarrow \text{false}\right]$.

Clause 1 Lower Interval Limit

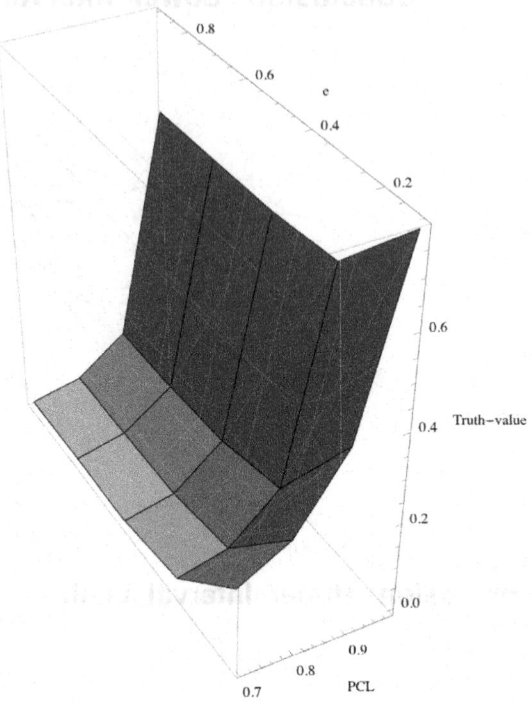

Clause 1 Upper Interval Limit

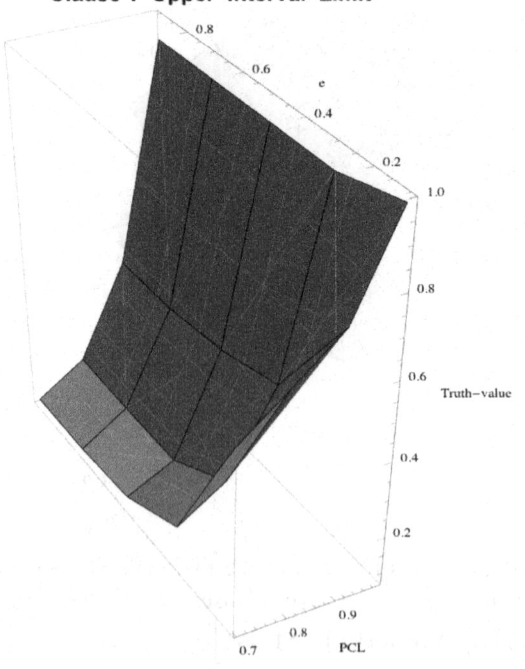

For clause 2,
$\left[\text{ThereExists } y \middle| \left[\text{ForAll } x \middle| \left(\text{Winner}(x) \parallel \text{Winner}(y)\right) \rightarrow \text{Winner}(y)\right]\right]$, all the lower limits are 0. The graph of the upper limit is:

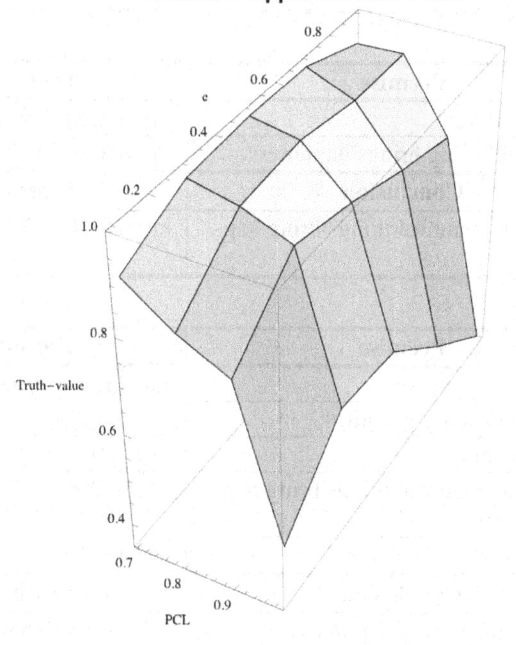

For our second example, we consider a simple probabilistic syllogism, but expressed using natural language quantifiers:

Many women are beautiful.
Almost all beautiful things bring happiness.
|-
Many women bring happiness

In order to use the indefinite probabilities formalism, we first need to determine appropriate values for the parameters LPC and UPC to represent the fuzzy concepts "many" and "almost all." In practice, in the case where these rules are used within an integrative AGI system such as the NCE, appropriate values for these fuzzy concepts will be determined by the context in which they appear. In one context, for example, the interval [0.8, 0.9] might represent the idea "many," but in a different situation we may wish for [0.6, 0.95] to represent "many."

For our example we set e=0.1. Let us suppose that "many" is represented by the interval [LPC, UPC]= [0.4, 0.95], and "almost all" by the interval [0.9, 0.99]. We will also assume truth-values identical to those in the previous example. The sequence of conclusions is then illustrated in the following tables.

Premise	**Truth-value**
Women	<[0.45, 0.55], 0.9, 10>
An individual woman is beautiful	<[0.8, 0.95], 0.9, 10>
Conclusion	**Truth-value**
Many women are beautiful	<[0.35451, 0.63574], 0.9, 10>

Premise	**Truth-value**
Beautiful things	<[0.4, 0.8], 0.9, 10>
A beautiful thing brings happiness	<[0.8, 0.95], 0.9, 10>
Conclusion	**Truth-value**
Almost all beautiful things bring happiness	<[0.03906, 0.37464], 0.9, 10>

Premise	**Truth-value**
Women	<[0.45, 0.55], 0.9, 10>
Many women are beautiful	<[0.35451, 0.63574], 0.9, 10>
Beautiful things	<[0.4, 0.8], 0.9, 10>
Almost all beautiful things bring happiness	<[0.03906, 0.37464], 0.9, 10>
Happiness	<[0.4, 0.9], 0.9, 10>
Conclusion	**Truth-value**
Many women bring happiness	<[0.41308, 0.53068], 0.9, 10>

Chapter 12: Intensional Inference

Abstract In this chapter we make precise the notion of "intensional inheritance," by which is meant inheritance based on "properties" or "patterns." We have mentioned intensional or mixed inheritance relationships here and there previously, but have not formally defined them.

12.1 Introduction

We have already sketched out the basic notion of intensional inheritance used in PLN: we define intensional inheritance by associating entities with pattern-sets, so that, e.g., fish would be associated with $fish_{PAT}$, the set of all patterns associated with fish. The intensional inheritance between fish and whale is then defined as the Subset relationship between $fish_{PAT}$ and $whale_{PAT}$, and the composite Inheritance between two entities is defined as the disjunction of the Subset and IntensionalInheritance between the entities. However, the concept of "pattern" used here was not formally defined previously; that is one of our tasks here.

Before getting into formal details however, the conceptual foundations of the intension/extension distinction are worth reviewing in more depth because they become somewhat subtle. Intuitively, for instance, we might say:

```
Subset whale fish <0>
IntensionalInheritance whale fish <.7>
```

Yet one might argue that the IntensionalInheritance relation is unnecessary here, because using only SubsetRelationships we could reason using PLN that

```
Subset fish (lives_in_water AND swims)
Subset whale (lives_in_water AND swims)
|-
Subset fish whale <tv>
```

where tv.s > 0.

But the problem is that this abductive inference would be based on the erroneous assumption that *fish* and whale *are* independent in *(lives_in_water AND swims)*. It would be overridden, for example, by textbook knowledge giving a definition of a fish as a complex predicate combining various attributes. What we want with intensional inheritance is some way of saying that *fish* inherits from *whale* that is NOT overridden by textbook knowledge. This is provided by the recourse to pattern-sets.

Inference on composite InheritanceRelationships, we contend, better captures the essence of human commonsense reasoning than reasoning on SubsetRelationships. When we say something like

```
"Life is a bowl of (half-rotten) cherries"
```

we're not merely doing Bayesian inference on the members of the sets "life" and "bowl of (half-rotten) cherries" – we're noting an inheritance on the association or pattern level, which has a strength greater than would be obtained via Bayesian inference applied in the standard extensional way. Philosophically, the issue is not that probability theory is wrong; the issue is that the most straightforward way of applying probability theory is not in accordance with common human practice, and that in this case common human practice has a good reason for being roughly the way it is (the reason being that finding similarities based on common patterns rather than common members is often useful for understanding a world in which many more patterns exist than would be expected at random).

12.2 Probabilistic Pattern Theory

Now we introduce the formal definition of the concept of *pattern* – a notion central to our analysis of intension but also highly significant in its own right. In Goertzel (1993, 1993a, 1997) a mathematical theory of pattern is outlined, based on algorithmic information theory. The conceptual foundation of that theory is the notion of a pattern as a "representation as something simpler." In Goertzel (2006), a modification to pattern theory is presented in which the algorithmic-information notions are replaced with probability-theoretic notions, but the conceptual spirit of the original pattern theory is preserved. Here we give a brief outline of probabilistic pattern theory, with PLN applications in mind.

12.2.1 Association

First, we introduce the notion of *association*. We will say that a predicate F is associated with another predicate G if

$$P(F|G) > P(F|\neg G)$$

That is, the presence of F is a positive indicator for the presence of G. The degree of association may be quantified as

$$\mathrm{ASSOC}(F,G) = \left[P(F|G) - P(F|\neg G)\right]^{+}$$

(where $[x]^+$ denotes the positive part of x, which is x if $x>0$ and 0 otherwise).

The association-structure of G may be defined as the fuzzy set of predicates F that are associated with G, where ASSOC is the fuzzy-set membership function. Of course, the same definition can be applied to concepts or individuals rather than predicates.

One may also construct the "causal association-structure" of a predicate G, which is defined the same way, except with temporal quantification. One introduces a time-lag T, and then defines

```
ASSOC(F,G; T) =
[P(F is true at time S+T | G is true at time T) -
P(F is true at time S+T|~G is true at time T)]+
```

The temporal version may be interpreted in terms of possible-worlds semantics; i.e.,

```
P(F at time S+T|G at time T)) = the probability that
the entity A will be satisfied in a random universe at
time A, given that G is satisfied in that universe at
time T-A

P(F) = the probability that predicate F will be sat-
isfied in a random universe, generically
```

If A and B are concepts or individuals rather than predicates, one can define ASSOC(A,B;T) by replacing each of A and B with an appropriately defined predicate. For instance, we can replace A with the predicate F_A defined by

```
ExtensionalEquivalence
    Evaluation FA U
    Subset A U
```

12.2.2 From Association to Pattern

From association to pattern is but a small step. A pattern in a predicate F is something that is associated with F, but is simpler than F. A pattern in an entity A is defined similarly.

We must assume there is some measure c() of complexity. Then we may define the fuzzy-set membership function called the "pattern-intensity", defined as, e.g.,

```
IN(F,G) = [c(G) - c(F)]+ [P(F | G) - P(F|~G)]+
```

There is also a temporal version, of course, defined as

```
IN(F,G;T) = [c(G;T) - c(F;T)]+ * ASSOC(F,G;T)
```

The definition of complexity referenced here can be made in several different ways. Harking back to traditional pattern theory, one can use algorithmic information theory, introducing some reference Turing machine and defining c(X) as the length in bits of X expressed on that reference machine. In this case, complexity is independent of time-lag T.

Or one can take a probabilistic approach to defining complexity by introducing some reference process H, and defining

```
c(F;T) = ASSOC(G,H;T)
```

In this case we may write

```
IN[H](F,G;T) = [ASSOC(F,H;T) - ASSOC(F,G;T)]+ *
ASSOC(F,G;T)
```

Notation-wise, typically we will suppress the H and T dependencies and just write IN(F,G) as a shorthand for IN[H](F,G;T), but of course this doesn't mean the parameters H and T are unimportant. Note the dependence upon the underlying computational model is generally suppressed in both traditional pattern theory and algorithmic information theory.

It's worth noting that this approach can essentially replicate the algorithmic information theory approach to complexity. Suppose that M is a system that spits out bit strings of length <=N – choosing each bit at random, one by one, and then stopping when it gets to a bit string that represents a self-delimiting program on a given reference Turing machine. Next, define the predicate H as being True at a given point in time only if the system M is present and operational at that point in time. Next, for each bit string B of length <=N consider the predicate F[B](U,T) that returns True if B is present in possible-universe U at time T and False otherwise. Then it is clear that P(F[B]) is effectively zero (in a universe without a program randomly spitting out bit strings, the chance of a random bit string occurring is low). On the other hand, -log(P(F[B] at time S | H at time S-T)) is roughly equal to the length of B. So for this special predicate H and the special predicates F[B], the complexity as defined above is basically equal to the negative logarithm of the program length. Algorithmic information theory emerges as a special case of the probabilistic notion of complexity.

We can introduce a notion of relative pattern here – which is useful if patterns need to be calculated contextually. If we assume K as background knowledge, then we can define

```
ASSOC(F,G|K) = [P(F | G & K) - P(F|~G & K)]+
```

and

```
ASSOC(F, G| K ; T) =
[P(F at time S+T | G & K at time T) -
P(F at time S+T|~G & K at time T)]+
```

and

```
IN(F,G|K;T) = [c(G|K;T) - c(F|K;T)]+ * ASSOC(F,G|K;T)
IN[H](F,G|K;T) = [ASSOC(F,H|K;T) - ASSOC(F,G|K;T)]+ *
ASSOC(F,G|K;T)
```

Using these ideas, associated with any concept or predicate, we may construct a set called the "structure" of that entity. We may do this in (at least) two ways:

- The associative structure, which consists of the set of all observations and concepts that are associated with the entity
- The pattern structure, which is the set of all observations and concepts that are patterns in the entity

For example, returning to our friend Stripedog, we may call these two sets *stripedog_ASSOC* and *stripedog_PAT*. Note that *stripedog_ASSOC* and *stripedog_PAT* are sets of observations or concepts, whereas *stripedog*, as defined above, is a set of sets of observations. *stripedog* is a fuzzy set (as well as a fuzzy cat), since the truth-value of the Evaluation relationship

```
Evaluation F_stripedog $X
```

need not be Boolean. And according to the definitions given above, *stripedog_ASSOC* is also a fuzzy set because it's not always entirely clear to what extent a given entity is associated with the individual Stripedog.

12.3 Intensional Inheritance Relationships

Now we will use the concepts of the previous section to give a novel perspective on the notion of *intensional inheritance*, as qualitatively introduced above. Suppose we want to compare our orange cat, Stripedog, to an orange (a particular fruit; let's name it orange_77). The set of entities *stripedog_PAT* associated with Stripedog includes observations such as

```
"orange is observed by me"
```

But this same observation is also included in *orange_77_PAT*. From this and other overlapping items, we may calculate that the probability associated with the relationship

```
ExtensionalSimilarity stripedog_PAT orange_77_PAT
```

is reasonably high.

In this way, we arrive as a similarity between an orange cat and an orange fruit. Note that no individual orange cats are orange fruits. Nor do any sets of observations need to be identifiers for both orange cats and orange fruits. Rather, all that's observed in the above similarity relationship is that there exists an overlap between the set of observations corresponding to the orange cat and the set of observations or properties corresponding to the orange fruit. So based on the above we cannot say

```
ExtensionalSimilarity stripedog orange_77
```

In fact this latter relationship may have a probability of zero.

As noted earlier, a similar effect could be obtained via use of abductive inference. If one observes that

```
Subset stripedog orange
Subset orange_77 orange
```

and then uses Bayes' rule to calculate

```
Subset orange orange_77
```

and then uses PLN deduction to calculate

```
Subset stripedog orange
Subset orange orange_77
|-
Subset stripedog orange_77
```

one will obtain a nonzero truth-value for the conclusion. However, this inference is based upon the erroneous assumption that stripedog and orange_77 are independent inside the set "orange." While it is nice that erroneous assumptions can sometimes lead to useful results, this is not a generally sound basis for commonsense inference. In practice it will lead to much lower strength values than humans typically obtain, because of the low strength values frequently produced by Bayes' rule. We feel that looking at pattern and association sets provides a better qualitative model of human commonsense inference in many cases – a perspective to be elaborated below.

Based on this conceptual and mathematical line of thinking, there are several ways to structure a pattern-based reasoning system. One approach would be to construct separate terms for C, C_ASSOC and C_PAT, corresponding to each individual. Sets of individuals, such as the set of cats, would also spawn such distinct terms; there would be a *cat* term defined as the set of individual cats, and *cat_ASSOC* and *cat_PAT* terms defined as the set of observations associated with individual cats. Consistently with the above, the membership function of *cat_ASSOC* would be defined as

$$[\ P(cat|\ E) - P(cat|\sim E)\ ]^+$$

where P(cat) means P(some element of the set of cats is observed), which is equivalent to P(some identifier of some cat is observed). The membership function of cat_PAT would be defined similarly as

$$[c(cat) - c(E)]^+ * [\ P(cat|\ E) - P(cat|\sim E)\ ]^+$$

However, this is not the implementation we have chosen. We believe this duplication of terms would be conceptually awkward – it doesn't reflect the way human commonsense reasoning operates. Rather, we humans seem to fairly freely mix up reasoning about patterns with reasoning about sets of individuals. PLN is designed to enable this mixing-up, while at the same time allowing a distinction to be made clearly when this is appropriate.

In order to do this, we introduce the notion of an IntensionalInheritance relationship. Subset, as defined above, has a truth-value defined simply by conditional probability. IntensionalInheritance, on the other hand, is a special relationship defined by the convention that, e.g.,

```
IntensionalInheritance stripedog orange_77
```

means

```
Subset stripedog_ASSOC orange_77_ASSOC
```

Formally, we may introduce the ASSOC operator, defined as

```
ExtensionalEquivalence
    Member $X (ExOut ASSOC $C)
    AND
          ExtensionalImplication
                Subset $Y $E
```

```
                    Subset $Y $C
          NOT
                    ExtensionalImplication
                              Subset $Y $E
                              NOT [Subset $Y $C]
```

and then define the associational version of IntensionalInheritance via the relationship

```
ExtensionalEquivalence
    IntensionalInheritance_ASSOC $X $Y
    Subset (ExOut ASSOC $X) (ExOut ASSOC $Y)
```

Next, we may define the ASSOC variant of the Inheritance relationship via

```
ExtensionalEquivalence
    Inheritance_ASSOC $X $Y
    OR
          Subset $X $Y
          IntensionalInheritance_ASSOC $X $Y
```

And, analogously to the above, we may define a symmetric mixed intensional/extensional relationship,

```
Similarity_ASSOC A B <tv>
=
OR
          ExtSim A B <tv1>
          ExtSim A_ASSOC B_ASSOC <tv2>
```

Of course, one can rewrite the above definitions using PAT instead of ASSOC. On the other hand, if we want to use both association-structure and pattern-structure, then in fact things get a little more complicated. We have three different kinds of inheritance to be disjunctively combined; e.g.,

```
ExtensionalEquivalence
    Inheritance_ASSOC_PAT $X $Y
    OR
          Subset $X $Y
          IntensionalInheritance_ASSOC $X $Y
          IntensionalInheritance_PAT $X $Y
```

Typically we will leave off the qualifiers and just refer to Inheritance rather than Inheritance_ASSOC, etc.

Note a significant difference from NARS here. In NARS, it is assumed that X inherits from Y if X extensionally inherits from Y but Y intensionally inherits from (inherits properties from) X. We take a different approach here. We say that X inherits from Y if X's members are members of Y, and the observations associated with X are also associated with Y. The duality of properties and members is taken to be provided via Bayes' rule, where appropriate.

By making frequent use of this kind of Inheritance relationship in PLN we are making the philosophical assertion that commonsense inference habitually mixes up extensional and intensional inheritance, as defined here, in a disjunctive way. Clearly this particular way of combining probabilities is not mathematically "necessary" or natural in the sense that conditional probabilities are. However, we contend that it is *cognitively* natural.

We believe that it is very cognitively valuable to calculate intensional inheritance and similarity values directly based on the above formulas – and use these together with extensional inheritance and similarity values within PLN inference. This is not hard to do in practice – it's merely a matter of arranging conditional probability calculations in a different way; a way motivated by human psychology and pattern theory rather than probability theory proper. It is a credit to the flexibility of the PLN framework that it supports this kind of variation so easily.

We should note that the calculation of

```
Intensional Inheritance A B
```

involves a heuristic approximation because the probabilities inside, for instance, $\chi_{ASSOC(A)}C_1$ and $\chi_{ASSOC(A)}C_2$, where χ represents the fuzzy membership function (and $\chi_{ASSOC(\$C)}\$X$ is simply a different notation for the truth value of `Member $X (ExOut ASSOC $C))` are not, in general, independent for the two properties C_1 and C_2.

One way to minimize this sort of independence-assumption-based error is to choose a set of properties that are relatively independent of each other -- a sort of approximately-orthogonal ontology spanning concept-space. This same sort of approximately-orthogonal decomposition may also be useful in default inference which we discuss in chapter 13.

12.3.1 Intensional Term Probabilities

Next, we touch on some mathematical issues raised by these constructions. Since we're asking a term like *stripedog* to serve as a proxy for two or more terms – *stripedog* and *stripedog_ASSOC* and *stripedog_PAT* – it would seem that the term should have several, separate term probabilities:

- An extensional term probability, indicating the extent to which a random observation-set is an identifier for *stripedog* and counts as an observation of *stripedog*
- Intensional term probabilities, indicating the extent to which a random observation is associated with, or is a pattern in, *stripedog*

We can define a composite term truth-value via

$$P_{Inheritance}(\$X) = OR(\ P(\$X),\ P_{ASS}(\$X),\ P_{PAT}(\$X))$$

Obviously these Inheritance term-probabilities go naturally with Inheritance relationships, whereas the purely extensional term probabilities go naturally with Subset relationships.

12.4 Truth-value Conversion Formulas

Next, the question arises how to convert between the extensional and mixed versions of inheritance relationships. Taking the ASSOC version of Inheritance for simplicity, the conversion rule

```
Inheritance A B
|-
Subset A B
```

has the form

```
OR(X,Y)
|-
Y
```

and the conversion rule

```
Subset A B
|-
Inheritance A B
```

has the form

```
X
|-
X OR Y
```

Obviously these are badly underdetermined inference rules, and the only solutions are simple heuristic ones. In the first case, we use the heuristic that

```
s = (OR(X,Y).tv.s
Y.tv.s = (s-c)/(1-c)
```

where c is a constant that may be set equal to the average strength of all IntensionalInheritance relations in the system's memory. In the second case we use the heuristic that

```
s = X.tv.s
 (OR(X,Y)).tv.s = c + (1-c)s
```

where c is defined the same way.

12.5 Intensional Boolean Operators

Next, we turn to the intensional versions of the standard Boolean operators discussed in the previous chapter. In general, in PLN we may consider Boolean operators to act either purely extensionally, purely intensionally, or in a mixed intensional/extensional manner. For instance, one may consider (A AND B) purely extensionally, as was done above, by taking the intersection of the sets of members of A and B. This is the default and the simplest approach, but not always the best route. Or, one may also consider (A AND B) purely intensionally, by taking the intersection of the sets of patterns associated with A and B. In this way, along with the standard (extensional) AND, OR, and NOT, we may define operators such as AND_{Int} and OR_{Int}.

And, as it happens, certain combinations of intensional and extensional Boolean operations are more useful than others. For instance, it generally makes sense to pair extensional intersection with intensional union: the members of the intersection of A and B will generally share the properties of both A and B. Thus we may define (A AND_{Mix} B) as the term containing the intersection of the extensional relationships of A and B, and the union of the intensional relationships of A and B. Similarly, (A OR_{Mix} B) is the term containing the union of the extensional relationships of A and B, and the intersection of the intensional relationships of A and B.

These various operators may be useful for generating new concepts to be utilized in future inferences. For instance, (cat AND_{Int} dog) is a category of hypothetical creatures sharing the properties of both cats and dogs. On the other hand, both (cat AND_{Mix} pet) and (cat AND_{Ext} pet) are the interpretations of the informal concept "the set of pet cats" but may be different fuzzy sets, because both cat and pet may have intensional relationships that were not derived from analysis of the members of these sets (but were derived, perhaps, from natural language interpretation). In (cat AND_{Mix} pet) these properties are carried over to the intersection but in (cat AND_{Ext} pet) they are not.

Apart from these logical operators, it may also be interesting to create new terms by simply taking the union or intersection of all relationships belonging to two terms A and B. However, it is important to understand that this kind of simplistic intersection and union is not the most conceptually natural approach in terms of the logic of either intension or extension.

12.6 Doing Inference on Inheritance Relationships

Having defined generic Inheritance as a disjunction of extensional and intensional inheritance, it is now natural to ask whether there are special inference rules for dealing with intensional and generic Inheritance relationships.

For instance, suppose we have

```
IntensionalInheritance A B
IntensionalInheritance B C
```

Can we then conclude

```
IntensionalInheritance A C
```

via "intensional deduction"? In fact, it seems that the best way to deal with this kind of situation is to use the definition of intensional inheritance in terms of probabilities. From

```
IntensionalInheritance A B
```

along with P(A) and P(B), one can figure out P(A|B); and similarly from

```
IntensionalInheritance B C
```

one can figure out P(C|B). Using ordinary (extensional) deduction one can then figure out P(C|A), and then from this plus P(C) and P(A) one can calculate the strength of

```
IntensionalInheritance A C
```

There is thus no need for a special "intensional deduction rule," and the same basic method holds for other inference rules. What about mixed inference? Suppose we have

```
Inheritance A B
```

```
Inheritance B C
```

Can

```
Inheritance A C
```

be concluded? Again, the story is a bit more complicated. The best thing is to record information separately about extensional and intensional inheritance. We can then combine them as weighted combinations of the following four types of inference:

```
1)    ExtensionalInheritance A B
      ExtensionalInheritance B C
      |-
      ExtensionalInheritance A C

2)    IntensionalInheritance A B
      IntensionalInheriatnce B C
      |-
      IntensionalInheritance A C

3)    ExtensionalInheritance A B
      IntesionalInheritance B C
      |-
      Inheritance A C (mixed Extensional/Intensional)

4)    IntensionalInheritance A B
      ExtensionalInheritance B C
      Inheritance B C (mixed)
```

On the other hand, sometimes one has information only about inheritance rather than about extensional and intensional inheritance particularly. This may happen, for instance, if one is dealing with knowledge parsed from natural language. The sentence "cats are animals" is best interpreted in terms of mixed inheritance. So if we are doing reasoning of the form

```
cats are animals
animals are ugly
|-
cats are ugly
```

then the best course is to map each of the premises into mixed inheritance relationships, then interpret each mixed inheritance relationship as a weighted sum of an extensional inheritance relationship and an intensional inheritance relationship. Inference may then be done to determine the truth-values of

```
IntensionalInheritance cat ugly
```

and

```
ExtensionalInheritance cat ugly
```

and these may be combined to get the truth-value of

```
Inheritance cat ugly
```

There is significant arbitrariness here in the determination of the weighting factor. It may be set arbitrarily as a system parameter, or else it may be determined "inferentially" via analogy to cases where the extensional and intensional inheritance values are known. For instance, if it is known that

```
ExtensionalInheritance dog ugly <.41,.3>
IntensionalInheritance dog ugly <.38,.5>
ExtensionalInheritance hog ugly <.79,.2>
IntensionalInheritance hog ugly <.82,.6>
```

then this may be taken as evidence that for cats as well,

```
ExtensionalInheritance cat ugly
```

should get a lower weighting than

```
IntensionalInheritance cat ugly
```

In all, then, intensional inference in PLN emerges as a different way of organizing the same basic probabilistic information that is used in extensional inference. The intension vs. extension distinction is important in terms of philosophy and cognitive science, yet in terms of the mathematics of probabilistic inference is merely a minor variation in the utilization of the familiar extensional inference rules.

12.7 Doing Inference on Inheritance Relationships

So far in our discussion of intension we have considered only strength values, and have not looked at weight of evidence or indefinite probabilities at all. Now we remedy this omission and explore how the indefinite probabilities framework may be extended to handle intensional inference. As usual in PLN, the tricky part is getting the semantics right; the mathematics follows fairly directly from that.

Consider the link

```
IntensionalInheritance whale fish
```

Suppose we want to assign an indefinite probability to this link (via direct evaluation rather than inference, initially): what's the semantics? Crudely, the semantics is

```
P( x in PROP(fish) | x in PROP(whale) )
```

where PROP(A) may be defined as ASSOC(A) or PAT(A), depending on which variant one is using. One way to conceive of PROP(A), in either case, is as a fuzzy set to which B belongs to a degree determined by a specific formula such as

```
χA(B) = floor[(P(A | B) - P(A | ~B) ) (1 - |A| / |B|)
]
```

in the case PROP=PAT.

If conditional probabilities such as P(A|B) are characterized by second-order probability distributions (as in the indefinite probabilities approach), then as a consequence $\chi_A(B)$ will be characterized as a second-order probability distribution over the space of fuzzy set membership values. For instance, suppose we choose a specific joint distribution for (P(A|B), P(A|~B)); then this will give rise to a specific distribution for $\chi_A(B)$. Choosing various joint distributions for (P(A|B), P(A|~B)), according to a second-order probability distribution specified for these conditional probabilities, one gets various distributions for $\chi_A(B)$; so one can then derive an envelope of distributions for $\chi_A(B)$. (Of course, we can choose these joint distributions using an independence assumptions if we want to; e.g., if we lack the information to do otherwise.) What this means is that the fuzzy membership value $\chi_A(B)$ is not a single value, but is rather an envelope of distributions describing a value, which may be summarized in many ways; for example, as an indefinite probability.

What we need to know to evaluate the truth-value of the above IntensionalInheritance relationship, then, is: Given the knowledge that X is a property of whale in the above sense, what's the chance that X is also a property of fish? Obviously this can be quantified in a number of ways. One appealing way involves the choice function

```
fA(B) = [ E(χA(B)) * woe(χA(B))]
```

where E(χA(B)) tells you the expected value of the envelope of distributions corresponding to $E(\chi_A(B))$, and $woe(\chi_A(B))$ tells you the weight of evidence calculated from this envelope of distributions (either approximated as the b-level confidence interval width of the set of means of distributions in the envelope for some fixed b, or else calculated more elaborately using assumptions or knowledge about the underlying distributions).

To calculate the truth-value of the above link, then, we can repeatedly carry out the following series of steps:

1. Choose a property R of whale with a probability proportional to f_R(whale), where f is the default choice function specified above.
2. Estimate $E(\chi_R(\text{fish}))$.

Doing this repeatedly gives us a distribution of means, which we can summarize using an indefinite probability. All this specifies how to define the indefinite probability governing the degree to which whale inherits from fish according to direct evaluation. If for some reason we want the whole envelope of distributions corresponding to χ_R(fish), then we simply have to retain this envelope in Step 2 above.

To calculate IntensionalSimilarity, on the other hand, one may use a minor variation on the above methodology. Continuing with the whale/fish case, for example, one can repeatedly carry out the following series of steps:

1. Choose a random pair (R, Z), where Z is either whale or fish and R is a property of one of them, and the probability of choosing the pair is proportional to $f_R(Z)$, where f is the default choice function specified above.
2. Estimate $E(\chi_R(W))$, where W is either whale or fish and is different from Z.

Doing this repeatedly gives us a distribution of means, which we can summarize using an indefinite probability.

Chapter 13: Aspects of Inference Control

Abstract In this chapter we will describe a few important aspects of the mathematical and software structures via which PLN has been implemented within the Novamente Cognition Engine – an implementation that has been structured with the goal of allowing PLN to serve the broadest possible variety of functions within an integrative AI context.

13.1 Introduction

The focus of this book is PLN mathematics, not software engineering, cognitive science, or integrative AI. However, the purpose with which the PLN mathematics has been conceived is closely tied to these other two topics; we have sought to create an uncertain inference framework capable of serving as a pragmatic uncertain inference component within the integrative Novamente AI system. For simplicity we will avoid giving any details of the Novamente architecture, presenting only the PLN implementation architecture itself. The mathematical, software, and conceptual relationships between PLN and the remainder of the NCE are beyond the scope of this book.

13.2 Forward- and Backward-Chaining Inference

The PLN inference rules and formulas are local in character; they tell you how to produce a conclusion from a particular set of premises. They don't address what is in a sense the most difficult aspect of inference: the determination of which inference rules to perform in what order. We call this process "inference control." Inference control in PLN is handled via mechanisms that are familiar from traditional AI – forward and backward chaining – but with special twists enforced by the omnipresence of uncertainty in PLN inference.

Generically, in PLN or elsewhere, backward chaining is defined as an inference process that starts with one or more inference targets and works backward to find inference paths leading to these targets as conclusions. An inference engine using backward chaining starts by searching the available inference rules until it finds at has a conclusion of a form that matches a desired goal. It then seeks to find suitable premises for the rule – premises that will cause the rule to output the goal as a conclusion. In conventional, crisp-logic Boolean backward chaining, if appropriate premises for that inference rule are not known to be true, then they are added to the list of goals, and the process continues until a path is found in which all goals on the path have been achieved. In PLN backward chaining, the probabilistic truth-values necessitate that the appropriate premises for that inference rule are recorded as such but also added to the list of goals. The process continues until the target atom can be produced with truth-value weight higher than a threshold set by the controller. The trick of making this work effectively is "pruning": intelligently and adaptively choosing which rules to choose and which premises to feed them, iteratively as the process proceeds.

Generic forward chaining, on the other hand, starts with the available knowledge and uses inference rules to derive a series of conclusions. An inference engine using forward chaining searches the inference rules until it finds one whose inputs have the same form as the given premises, then uses the rules to produce output. The output is then added to the set of possible premises to be used in future inference steps. The tricky part here is once again pruning; in most cases there are many different rules from which to select, and it's not clear which are the right ones.

The bugaboo of both forward- and backward-chaining inference control is combinatorial explosion. We describe here some heuristics used to palliate this problem in PLN, but these heuristics do not solve the problem fully adequately, except for the case of fairly simple inferences. Of course, the problem of combinatorial explosion is not fully solvable without introducing implausibly massive computational resources, but it must be solved more thoroughly than is done here if PLN is to be useful for powerful general intelligence. In the overall Novamente architecture, the combinatorial explosion is intended to be quelled via the integration of PLN with other AI components such as evolutionary learning, stochastic pattern mining and adaptive attention allocation, topics that will not be discussed here.

13.3 Differences Between Crisp and Probabilistic Theorem-Proving

While the basic structures of PLN inference control (forward and backward chaining) are familiar from crisp-logic-based theorem-proving, significant differences arise due to the prominent role that uncertainty plays in PLN inference. In the automated proof of crisp logical theorems, the goal is to build an inference trail that proceeds, in each step, from premise predicates with crisp truth-values to new predicates with crisp truth-values. The goal in backward chaining, for example, is to find a single path of this nature leading from some set of known pre to the given target predicate.

In probabilistic theorem-proving, on the other hand, finding a single path from known premises to the target is not necessarily very useful, because the truth-value estimate achieved via this path may have low confidence. Multiple paths may yield multiple truth-value estimates, which must then be revised or chosen between. These complexities make inference control subtler, and among other factors they imply a more prominent role for "effort management." It is often easy to find *some* path from premises to the target if we have no requirements for the resulting truth-value. Therefore, in PLN backward chaining we are in practice more interested in *how many resources we are willing to allocate for improving the truth-value of the target by alternative inference paths,* rather than aiming for the first possible path we can find.

13.3.1 Atom Structure Templates

In order to carry out an individual step in backward- or forward-chaining inference, we need to be able to quickly figure out which rules can be applied to a certain Atom (for forward chaining) or which rules can yield a certain Atom as output (for backward chaining). In order to be able to do this conveniently, we need a way to describe Atom properties in a concise way. We will do this with predicates called *Atom structure templates.*

Given an Atom A, we may construct the predicate A^ so that A^(X) is true iff there is some Y so that A(Y)=X. In the case that A is a first-order Atom (a Term or a Relationship between Terms, with no variables or higher-order functions involved), then A^(X) is true iff X=A; but this is not the most interesting case.

E.g., for the Atoms

```
P =
Inheritance A B
```

and

```
Q =
Inheritance
    A
    $1
```

it is easy to see that Q^(P) is true (via the assignment $1=B) but P^(Q) is not.

We may also define more complex Atom types such as

```
HasTruthValueStrengthGreaterThan
```

```
        $1
        0.5
```

which can likewise be used as literals like any other Atoms, or as complex Atom structure templates that map Atoms into Boolean truth-values according to a specific property, such as "whether the truth-value.strength of the argument Atom is greater than 0.5."

Atom structure templates can also be used in conjunctions or disjunctions with other Atom structure templates. For example, the Atom structure template corresponding to

```
AND
    HasTruthValueStrengthGreaterThan
          $1
          0.5
    Inheritance
          A
          $1
```

is true of an Atom such that A inherits from it and its truth-value.strength > 0.5.

13.3.2 Rules and Filters

Recall the distinction between rules and formulas in PLN. A Formula is an object that maps a vector of truth-values into a single truth-value. For example, the deduction formula takes five truth-values and outputs the truth-value of the resulting Inheritance or Implication relationship. On the other hand, a Rule is an object that maps a vector of Atoms into a single new Atom. Along the way it applies a designated formula to calculate the truth-values. For example, the deduction rule takes two Inheritance relationships (A,B) and (B,C) and outputs a single Inheritance relationship (A,C) with a truth-value calculated by the deduction formula.

In the PLN software implementation, inference rules are represented by Rule objects. Each Rule object has both an input filter and an output-based-input filter, which will be used by forward chaining and backward chaining, respectively. The input filter is a vector of Atom structure templates that accept only specific kinds of argument Atoms in the respective place. For example, the input filter of DeductionRule is the vector of two Atoms representing Atom structure templates:

```
InheritanceLink
    $1
    $2
InheritanceLink
    $2
```

```
$3.
```

The output-based-input filter is more complex. It is used to detemine which Atoms are needed as parameters for the Rule in order to produce Atoms of desired kind. For example, in order to produce the

```
InheritanceLink
    A
    C
```

by the DeductionRule you need to provide, as input to the DeductionRule, the Atoms that satisfy the Atom structure templates

```
InheritanceLink
    A
    $1

InheritanceLink
    $1
    C
```

Of course a specific target may not be producible by a given rule, in which case the procedure simply returns a null result. In order to produce two Atoms that satisfy each one of the last two Atom structure templates, we need to begin a new proof process for each one separately, and so on. Since this particular proof contains Atom structure template variables, we must at some point unify the variables. Because they occur inside distinct Atoms, it may happen that a specific unification choice in one will prevent the production of the other altogether. We must then backtrack to this unification step, choose differently, and retry.

Note that producing InheritanceLink($1, C) is very different from producing an Atom that satisfies the Atom structure template InheritanceLink($1,C)^. Typically, in backward chaining we are interested in the latter.

13.4 Simplification

Some computational reasoning systems require all logical formulas to fall into a certain standard "normal form," but this is not the case in PLN. There is the option for normalization, and we plan to experiment with using Holman's Elegant Normal Form (Holman 1990) in PLN, as we do in other aspects of the NCE, but normalization of this nature is not required. As a default, the PLN implementation does only some elementary normalizations, which are not profoundly necessary for PLN inference control but do serve to simplify things and reduce the number of Rule objects required.

me "explicit" normalization happens in the proxy that the new Atoms pass before entering the knowledge base, and then some "implicit" normalization happens in the structures used to manipulate the Atoms. An example of the former kind of normalization is conversion of EquivalenceLinks into two ImplicationLinks, while an example of the latter kind is the mechanism that considers identical an Atom A and a single-member ListLink whose outgoing set consists of A.

13.5 Backward Chaining

We now describe the PLN backward-chaining inference process. Each step of this process is thought of as involving two substeps: expansion and evaluation. In the expansion step, the filters of Rule objects are used to figure out the possible child nodes for the current BIT (BackInferenceTree) node. These nodes are then inserted into an "expansion pool," which contains all the nodes that can still be expanded to find additional proof paths. In the evaluation step the BIT is traversed from leaves towards the root, where children of a node are presented as input arguments to the Rule object that the parent node represents. Because each child is in general a set of possible solutions for the given Rule argument, we have multiple combinations of solutions to try out.

As noted above, a central characteristic of PLN backward chaining is the fact that "topological" inference path-finding is only a subgoal of the main goal of PLN backward inference: maximizing the weight-of-evidence of the truth-value of the target Atom. So, in order to reach the latter goal, we may need to construct several different inference trees, with inference paths that may have considerable "topological" overlap yet assign different truth-values even to the overlapping nodes.

This process of multiple-tree generation can conveniently be described in evolutionary terms, where the genotype of a Backward Inference Tree (BIT) consists of the individual Atoms and the Rules that combine them, while the weight of evidence of the result produced by the tree is its fitness. To calculate the fitness of a BIT it is necessary to evaluate all the proof pathways in the tree, and then combine the final query results using the Rule of Choice.

On the implementation level there is some subtlety involved in constructing the trees involved in backward-chaining inference, because the validity of a result may depend on other results that come from other branches. A whole new search tree must be created every time the production of a target Atom is attempted. To achieve this, on each attempt to produce a target Atom we save the partial tree produced thus far, and launch a new tree that contains only a pointer to the partial tree. Thus, the task of improving the truth-value of an Atom requires partial or complete reconstruction of the proof tree several times. Therefore, a critical implementation requirement is the efficiency of the "retry mechanism" that allows for this.

Managing variable bindings in this tree expansion process is also tricky. I approach, to carry out backward-chaining inference, you expand the inference tree step-by-step, and you may evaluate the fitness of the tree after any step. To produce each subresult several Rules can be invoked, causing bifurcation. At the time of evaluation, some subresults are in fact sets of results obtained via different pathways. Hence, the evaluation step must look at all the consistent conbinations of these; consistency here meaning that bindings of the subresults used in a specific combination must not conflict with each other. Ideally, one might want to check for binding consistency at the expansion stage and simply discard conflicting results. However, in case of conflict we do not in general know which ones of the conflicting bindings are worse (i.e., will contribute to lower ultimate fitness) than others. Hence, it is best to do the consistency check at the evaluation stage.

13.6 Simple BIT Pruning

For pruning of the backward-chaining inference tree, we use initially a technique based on the multi-armed bandit problem from statistics (Robbins 1952; Sutton, Bartow 1998.)

Consider first a relatively simple case: the backward-chainer wants to expand a given node N of the BIT it is creating, and there is a number of possibilities for expansion, corresponding to a number of different rules that may possibly give rise to the expression contained at node N. Some of these have not yet been tried, but these have a priori probabilities, which may be set based on the general usefulness of the corresponding rule, or in a more sophisticated system based on the empirical usefulness of the rule in related contexts. Some may have already been expanded before, yielding a truth-value estimate with a known strength and weight of evidence.

Then, in our current inference control approach, each possibility for expansion (each rule) must be assigned an "expected utility" number indicating the probability distribution governing the degree of fitness increase expected to ensue from exploring this possibility. For unexplored possibilities this utility number must be the a priori probability of the rule in that context. For possibilities already extensively explored, this utility number may be calculated in terms of the weight of evidence of the truth-value estimate obtained via exploring this possibility, and the amount of computational effort expended in calculating this truth-value.

Regarding unexplored or lightly-explored possibilities, the two heuristics used to set the "a priori" utilities for BIT nodes are:

1. Solution space size: Prefer nodes with more restrictive proof targets (i.e., ones with fewer free variables and more structure).
2. Rule priority: Prefer e.g. direct atom lookup to a purely hypothetical guess.

take into account information gained by prior exploration of a BIT node, one may utilize a simple heuristic formula such as

```
u = d/E = weight_of_evidence / effort
```

or more generally

```
u = = F/E = fitness / effort
```

(where e.g. fitness might be measured as weight_of_evidence*strength). A more sophisticated variant might weight these two factors, yielding e.g.

$$u = F^a/E^b$$

Given these probabilities, an avenue for expanding the node may be selected via any one of many standard algorithms for handling multi-armed bandit problems; for initial experimentation we have chosen the SoftMax algorithm (Sutton, Bartow 1998), which simply chooses an avenue to explore at random, with a probability proportional to

$$e^{u/T}$$

where T is a temperature parameter. High temperatures make all possibilities equiprobable whereas T=0 causes the method to emulate greedy search (via always following the most probable path).

One of the strengths of this overall approach to inference control is its capability to incorporate, into each inference step, information gained via other sorts of inference or via non-inferential processes. This is particularly valuable when considering PLN as a component of an integrative AI system. For instance, if encounters the link

```
ExtensionalInheritance A B
```

as a possibility for exploration in a BIT, then a reasonable heuristic is to check for the existence and truth value of the link h

```
IntensionalInheritance A B
```

If A and B share a lot of properties, they may heuristically be guessed to share a lot of members; etc. Or, if there is no existing information directly pertinent to the ExtensionalInheritance relation being considered, and there are sufficient computational resources available, it might be worthwhile to execute some other non-inferential cognitive process aimed at gathering more relevant information; for instance, if many links of the form

```
MemberLink Ci B
```

are available, an evolutionary learning or other supervised categorization algorithm could be used to learn a set of logical rules characterizing the members of B, which could then be used in the inference process to gain additional information potentially pertinent to estimating the utility of expanding

```
ExtensionalInheritance A B
```

13.7 Pruning BITs in the Context of Rules with Interdependent Arguments

The pruning situation becomes more complex when we consider rules such as Deduction, which take arguments that depend on each other. For Deduction to produce Implication(A,C), input arguments are of the form (Implication(A,$1), Implication($1,C)) where $1 is a variable. Naively, one might just want to simplify this by looking up only Atom pairs that simultaneously satisfy both requirements. This fails, however, because we must also be able to *produce* both arguments by our regular inference process. So for each argument a separate inference tree is created, and the consistency check must be postponed.

Expanding both branches independent of each other is wasteful, because once we find a good result for one of the arguments in that branch, e.g., binding $1 to B, we would like to bias the inference in the other branch toward being consistent with this; i.e., producing Implication(B,C).

One approach would be to launch a new separate proof process for this purpose, but this is difficult to implement because in general there are other things going on in the proof beside this single (e.g., deduction) step, and we would like to have only a single control path. In this hypothetical approach it would be unclear when exactly we should continue with this proof process and when with the other one(s).

A simpler approach is to create a clone of the parent of the current inference tree node, with the one difference that in this new node the said variable (B) has been bound. The new node is then added to the execution pool.

This may seem counter-intuitive because we do not restrict the possible proof paths, but simply add a new one. For example, suppose that in this example Implication($1,C) is still in the expansion pool, and the Implication(A,$1) is currently under expansion. The expansion produces result Implication(A,B). Then, the execution pool will contain both Implication($1,C) and Implication(B,C).

Note that the whole execution pool will typically not be exhausted, but we only expand a part of it until a sufficiently fit result has been found. Further, we have set the default utilities so that Implication(B,C) has a higher a priori utility than Implication($1,C) because these nodes are identical in all other respects, but the r has fewer free variables and therefore a smaller solution space. It follows that the proof for Implication(B,C) will be attempted before the more general Implication($1,C). Of course, if Implication(B,C) does not exist and has to be produced by inference, then the process may end up expanding Implication($1,C) before it has been able to produce Implication(B,C). But this is fine because Implication(B,C) was simply our first guess, and if the Atom does not exist we have no strong reason to believe that the binding $1=>B should be superior to other alternatives.

The bandit problem based approach mentioned above may still be used in the more complex case of rules like deduction with interdependent arguments, but the method of calculating u must be modified slightly. In this case, the probability assigned to a given possible expansion might (to cite one sensible formula out of many possible ones, and to use the "weight of evidence" fitness function purely for sake of exemplification) be set equal to

$$u = (d_1 d_2)^a / E^b$$

where d_1 is the weight of evidence of the truth-value output by the expansion node, and d_2 is the weight of evidence of the inference rule output produced using the output of the expansion node as input. For instance, if doing a deduction

```
Inheritance A B <[0.8, 0.9], 0.9, 10>
Inheritance B C <[0.7, 0.8], 0.9, 10>
|-
Inheritance A C <[0.618498, 0.744425], 0.9, 10>
```

then a particular way of producing

```
Inheritance A B
```

would be assigned d_1 based on the weight of evidence obtained for the latter link this way, and would be assigned d_2 based on the weight of evidence obtained for

```
Inheritance A C
```

this way.

13.8 Forward-Chaining Inference

Finally, we consider the problem of forward-chaining inference. This is very closely comparable to backward-chaining inference, with the exception that the method of calculating utility used inside the bandit problem calculation is different. One is no longer trying to choose a path likely to yield maximum increase of weight of evidence, but rather a path likely to yield this along with maximu terestingness" according to some specified criterion. For example, one definition of interestingness is simply deviation from probabilistic independence. Another, more general definition is the difference in the result obtained by evaluating the truth-value of the Atom with very little effort versus a large amount of effort. So one might have

$$u = I^c d^a / E^b$$

where I measures interestingness; or more general versions where e.g. the weight of evidence d is replaced with some other fitness function. The measurement of interestingness is a deep topic unto itself, which brings us beyond PLN proper; however, one PLN-related way to gauge the interestingness of an Atom, relative to a knowledge base, is to ask how different the Atom's actual truth value is, from the truth value that would be inferred for the Atom if it were removed and then had its truth value supplied by inference. This measures how much surprisingness the Atom's truth value contains, relative to the surrounding knowledge.

Of course, forward-chaining inference can also invoke backward-chaining inference as a subordinate process. Once it has happened upon an interesting Atom, it can then invoke backward chaining to more thoroughly evaluate the truth-value of this Atom via a variety of different paths.

Very shallow forward-chaining inference may be used as a method of "concept creation." For instance, there is value in a process that simply searches for interesting Boolean combinations of Atoms. This may be done by randomly selecting combinations of Atoms and using them as premises for a few steps of forward-chaining inference. These combinations will then remain in memory to be used in future forward- or backward-chaining inference processes.

13.9 Default Inference

Inference control contains a great number of subtleties, only a small minority of which has been considered here. Before leaving the topic we will consider one additional aspect, lying at the borderline between inference control, PLN inference proper, and cognitive science. This is the notion of "default inheritance," which plays a role in computational linguistics within the Word Grammar framework (Hudson 1984), and also plays a central role in various species of nonmonotonic "default logic" (Reiter, 1980; Delgrande and Schaub, 2003). The treatment of default inference in PLN exemplifies how, in the PLN framework, judicious inference control and intelligent integration of PLN with other structures may be used to achieve things that, in other logic frameworks, need to be handled via explicit extension of the logic.

To exemplify the notion of default inheritance, consider the case of penguins, which do not fly, although they are a subclass of birds, which do fly. When one ers a new type of penguin, say an Emperor penguin, one reasons initially that they do not fly – i.e., one reasons by reference to the new type's immediate parent in the ontological hierarchy, rather than its grandparent. In some logical inference frameworks, the notion of hierarchy is primary and default inheritance of this nature is wired in at the inference rule level. But this is not the case with PLN – in PLN, correct treatment of default inheritance must come indirectly out of other mechanisms. Fortunately, this can be achieved in a fairly simple and natural way.

Consider the two inferences (expressed informally, as we are presenting a conceptual discussion not yet formalized in PLN terms)

```
A)
penguin --> fly <0>
bird --> penguin <.02>
|-
bird --> fly

B)
penguin --> bird <1>
bird --> fly <.9>
|-
penguin --> fly
```

The correct behavior according to the default inheritance idea is that, in a system that already knows at least a moderate amount about the flight behavior of birds and penguins, inference A should be accepted but inference B should not. That is, evidence about penguins should be included in determining whether birds can fly – even if there is already some general knowledge about the flight behavior of birds in the system. But evidence about birds in general should not be included in estimating whether penguins can fly, if there is already at least a moderate level of knowledge that in fact penguins are atypical birds in regard to flight.

But how can the choice of A over B be motivated in terms of PLN theory? The essence of the answer is simple: in case B the independence assumption at the heart of the deduction rule is a bad one. Within the scope of birds, being a penguin and being a flier are not at all independent. On the other hand, looking at A, we see that within the scope of penguins, being a bird and being a flier are independent. So the reason B is ruled out is that if there is even a moderate amount of knowledge about the truth-value of (penguin --> fly), this gives a hint that applying the deduction rule's independence assumption in this case is badly wrong.

On the other hand, what if a mistake is made and the inference B is done anyway? In this case the outcome could be that the system erroneously increases its estimate of the strength of the statement that penguins can fly. On the other hand, the revision rule may come to the rescue here. If the prior strength of (penguin --> fly) is 0, and inference B yields a strength of .9 for the same proposition, then the special case of the revision rule that handles wildly different truth-value estimates may be triggered. If the 0 strength has much more confidence attached to i the .9, then they won't be merged together, because it will be assumed that the .9 is an observational or inference error. Either the .9 will be thrown out, or it will be provisionally held as an alternate, non-merged, low-confidence hypothesis, awaiting further validation or refutation.

What is more interesting, however, is to consider the implications of the default inference notion for inference control. It seems that the following may be a valuable inference control heuristic:

1. Arrange terms in a hierarchy; e.g., by finding a spanning DAG of the terms in a knowledge base, satisfying certain criteria (e.g., maximizing total strength*confidence within a fixed limitation on the number of links).
2. When reasoning about a term, first do deductive reasoning involving the term's immediate parents in the hierarchy, and then ascend the hierarchy, looking at each hierarchical level only at terms that were not visited at lower hierarchical levels.

This is precisely the "default reasoning" idea – but the key point is that in PLN it lives at the level of inference control, not inference rules or formulas. In PLN, default reasoning is a timesaving heuristic, not an elementary aspect of the logic itself. Rather, the practical viability of the default-reasoning inference-control heuristic is a consequence of various other elementary aspects of the logic, such as the ability to detect dependencies rendering the deduction rule inapplicable, and the way the revision rule deals with wildly disparate estimates.

Chapter 14: Temporal and Causal Inference

Abstract In this chapter we briefly consider the question of doing temporal and causal logic in PLN, and give an example involving the use of PLN to control a virtually embodied agent in a simulated environment.

14.1 Introduction

While not as subtle mathematically as HOI, temporal logic is an extremely important topic conceptually, as the vast majority of human commonsense reasoning involves reasoning about events as they exist in, interrelate throughout, and change, over time. We argue that, via elaboration of a probabilistic event calculus and a few related special relationship types, temporal logic can be reduced to standard PLN plus some special bookkeeping regarding time-distributions.

Causal inference, in our view, builds on temporal logic but involves other notions as well, thus introducing further subtlety. Here we will merely scratch the surface of the topic, outlining how notions of causality fit into the overall PLN framework.

14.2 A Probabilistic Event Calculus

The Event Calculus (Kowalski & Sergo, 1986; Miller & Shanahan, 1991), a descendant of Situation Calculus (McCarthy & Hayes, 1969), is perhaps the best-fleshed-out attempt to apply predicate logic to the task of reasoning about commonsense events. A recent book by Erik Mueller (2006) reviews the application of Event Calculus to the solution of a variety of "commonsense inference problems," defined as simplified abstractions of real-world situations.

This section briefly describes a variation of Event Calculus called Probabilistic Event Calculus, in which the strict implications from standard Event Calculus are replaced with probabilistic implications. Other changes are also introduced, such as a repositioning of events and actions in the basic event ontology, and the introduction of a simpler mechanism to avoid the use of circumscription for avoiding the "frame problem." These changes make it much easier to use Event Calculus

asoning within PLN. The ideas in this section will be followed up in the following one, which introduces specific PLN relationship types oriented toward temporal reasoning.

We suggest that the variant of event calculus presented here, as well as being easily compatible with PLN, also results in a more "cognitively natural" sort of Event Calculus than the usual variants, though of course this sort of claim is hard to substantiate rigorously and we will not pursue this line of argument extensively here, restricting ourselves to a few simple points. Essentially, these points elaborate in the temporal-inference context the general points made in the Introduction regarding the overall importance of probability theory in a logical inference context:

- There is growing evidence for probabilistic calculations in the human brain, whereas neural bases for crisp predicate logic and higher-order logic mechanisms like circumscription have never been identified even preliminarily.
- The use of probabilistic implications makes clearer how reasoning about events may interoperate with perceptions of events (given that perception is generally uncertain) and data mining of regularities from streams of perceptions of events (which also will generally produce regularities with uncertain truth-values).
- As will be discussed in the final subsection, the pragmatic resolution of the "frame problem" seems more straightforward using a probabilistic variant of Event Calculus, in which different events can have different levels of persistence.

14.2.1 A Simple Event Ontology

Probabilistic Event Calculus, as we define it here, involves the following categories of entity:

- events
    - fluents
- temporal predicates
    - holding
    - initiation
    - termination
    - persistence
- actions
- time distributions
    - time points
    - time intervals
    - general time distributions

A "time distribution" refers to a probability density over the time axis; i assigns a probability value to each interval of time. Time points are considered pragmatically as time distributions that are bump-shaped and supported on a small interval around their mean (true instantaneity being an unrealistic notion both psychologically and physically). Time intervals are considered as time distributions corresponding to characteristic functions of intervals.

The probabilistic predicates utilized begin with the functions:

```
hold(event)

initiate(event)

terminate(event)
```

These functions are assumed to map from events into probabilistic predicates whose inputs are time distributions, and whose outputs are probabilistic truth-values. The class of "events" may be considered pragmatically as the domain of these functions.

Based on these three basic functions, we may construct various other probabilistic predicates, such as

```
holdsAt(event, time point) =
    (hold (event))(time point)

initiatedAt(event,time point) =
    (initiate(event))(time point)

terminatedAt(event, time point) =
    (terminate(event))(time point)

holdsThroughout(event, time interval) =
    (hold (event))(time interval)

initiatedThroughout(event, time interval) =
    (initiate (event))(time interval)

terminatedThroughout(event, time interval) =
    (terminate (event))(time interval)

holdsSometimeIn(event, time interval)  =
    "There exists a time point t in time interval T
          so that holdsAt(E,t)"

initiatedSometimeIn(event, time interval) =
```

```
    "There exists a time point t in time interval T
            so that initiatedAt(E,t)"

  terminatedSometimeIn(event, time interval) =
    "There exists a time point t in time interval T
            so that terminatedAt(E,t)"
```

It may seem at first that the interval-based predicates could all be defined in terms of the time-point-based predicates using universal and existential quantification, but this isn't quite the case. Initiation and termination may sometimes be considered as processes occupying non-instantaneous stretches of time, so that a process initiating over an interval does not imply that the process initiates at each point within that interval.

Using the SatisfyingSet operator, we may also define some useful schemata corresponding to the above predicates. For example, we may define SS_InitiatedAt via

```
Equivalence
    Member $X (SatSet InitatedAt(event, *))
    ExOut SS_InitiatedAt(event) $X
```

which means that, for instance, SS_InitiatedAt(shaving_event_33) denotes the time at which the event shaving_event_33 was initiated. We will use a similar notation for schemata associated with other temporal predicates, below.

Next, there are various important properties that may be associated with events, for example persistence and continuity.

```
Persistent(event)

Continuous(event)

Increasing(event)

Decreasing(event)
```

which are (like hold, initiate, and terminate) functions outputting probabilistic predicates mapping time distributions into probabilistic truth-values.

Persistence indicates that the truth-value of an event can be expected to remain roughly constant over time from the point at which is initiated until the point at which the event is terminated. A "fluent" is then defined as an event that is persistent throughout its lifetime. Continuous, Increasing, and Decreasing apply to non-persistent events, and indicate that the truth-value of the event can be expected to {vary continuously, increase, or decrease} over time.

For example, to say that the event of clutching (e.g., the agent clutching a ball) is persistent involves the predicate (isPersistent(clutching))([-infinity,infinity]).

Note that this predicate may be persistent throughout all time even if it is no throughout all time; the property of persistence just says that once the event is initiated its truth-value remains roughly constant until it is terminated.

Other temporal predicates may be defined in terms of these. For example, an "action" may be defined as an initiation and termination of some event that is associated with some agent (which is different from the standard Event Calculus definition of action).

Next, there is also use for further derived constructs such as

```
initiates(action, event)
```

indicating that a certain action initiates a certain event or

```
done(event)
```

which is true at time-point t if the event terminated before t (i.e., before the support of the time distribution representing t).

Finally, it is worth noting that in a logic system like PLN the above predicates may be nested within markers indicating them as hypothetical knowledge. This enables Probabilistic Event Calculus to be utilized much like Situation Calculus (McCarthy, 1986), in which hypothetical events play a critical role.

14.2.2 The "Frame Problem"

A major problem in practical applications of Event Calculus is the "frame problem" (McCarthy, 1986; Mueller, 2006), which – as usually construed in AI – refers to the problem of giving AI reasoning systems implicit knowledge about which aspects of a real-world situation should be assumed not to change during a certain period of time. More generally, in philosophy the "frame problem" may be construed as the problem of how a rational mind should bound the set of beliefs to change when a certain action is performed. This section contains some brief conceptual comments on the frame problem and its relationship to Probabilistic Event Calculus and PLN in general.

For instance, if I tell you that I am in a room with a table in the center of it and four chairs around it, and then one of the chairs falls down, you will naturally assume that the three other chairs did not also fall down – and also that, for instance, the whole house didn't fall down as well (perhaps because of an earthquake). There are really two points here:

1. The assumption that, unless there is some special reason to believe otherwise, objects will generally stay where they are; this is an aspect of what is sometimes known as the "commonsense law of inertia" (Mueller, 2006).

> The fact that, even though the above assumption is often violated in reality, it is beneficial to assume it holds for the sake of making inference tractable. The inferential conclusions obtained may then be used, or not, in any particular case depending on whether the underlying assumptions apply there.

We can recognize the above as a special case of what we earlier called "default inference," adopting terminology from the nonmonotonic reasoning community. After discussing some of the particulars of the frame problem, we will return to the issue of its treatment in the default reasoning context.

The original strategy John McCarthy proposed for solving the frame problem (at least partially) was to introduce the formal-logical notion of circumscription (McCarthy, 1986). For example, if we know

```
initiatesDuring(chair falls down, T1)
```

regarding some time interval T1, then the circumscription of holdsDuring and T1 in this formula is

```
initiatesDuring(x,T1) <==> x = "chair falls down"
```

Basically this just is a fancy mathematical way of saying that no other events are initiated in this interval except the one event of the chair falling down. If multiple events are initiated during the interval, then one can circumscribe the combination of events, arriving at the assertion that no other events but the ones in the given set occur. This approach has known shortcomings, which have been worked around via various mechanisms including the simple addition of an axiom stating that events are by default persistent in the sense given above (see Reiter, 1991; Sandewall, 1998). Mueller (2006) uses circumscription together with work-arounds to avoid the problems classically found with it.

However, none of these mechanisms is really satisfactory. In a real-world scenario there are always various things happening; one can't simply assume that nothing else happens except a few key events one wants to reason about. Rather, a more pragmatic approach is to assume, for the purpose of doing an inference, that nothing important and unexpected happens that is directly relevant to the relationships one is reasoning about. Event persistence must be assumed probabilistically rather than crisply; and just as critically, it need be assumed only for appropriate properties of appropriate events that are known or suspected to be closely related to the events being reasoned about in a given act of reasoning.

This latter issue (the constraints on the assumption of persistence) is not adequately handled in most treatments of formal commonsense reasoning, because these treatments handle "toy domains" in which reasoning engines are fed small numbers of axioms and asked to reason upon them. This is quite different from the situation of an embodied agent that receives a massive stream of data from its sensors at nearly all times, and must define its own reasoning problems and its own relevant contexts. Thus, the real trickiness of the "frame problem" is not e what the logical-AI community has generally made it out to be; they have side-stepped the main problem due to their focus on toy problems.

Once a relevant context is identified, it is relatively straightforward for an AI reasoning system to say "Let us, for the sake of drawing a relatively straightforward inference, make the provisional assumption that all events of appropriate category in this context (e.g., perhaps: all events involving spatial location of inanimate household objects) are persistent unless specified otherwise." Information about persistence doesn't have to be explicitly articulated about each relevant object in the context, any more than an AI system needs to explicitly record the knowledge that each human has legs – it can derive that Ben has legs from the fact that most humans have legs; and it can derive that Ben's refrigerator is stationary from the fact that most household objects are stationary. The hard part is actually identifying the relevant context, and understanding the relevant categories (e.g., refrigerators don't move around much, but people do). This must be done inductively; e.g., by knowledge of what contexts have been useful for similar inferences in the past. This is the crux of the frame problem:

- Understanding what sorts of properties of what sorts of objects tend to be persistent in what contexts (i.e., learning specific empirical probabilistic patterns regarding the Persistent predicate mentioned above)
- Understanding what is a natural context to use for modeling persistence, in the context of a particular inference (e.g., if reasoning about what happens indoors, one can ignore the out of doors even it's just a few feet away through the wall, because interactions between the indoors and out of doors occur only infrequently)

And this, it seems, is just plain old "AI inference" – not necessarily easy, but without any obvious specialness related to the temporal nature of the content material. As noted earlier, the main challenge with this sort of inference is making it efficient, which may be done, for instance, by use of a domain-appropriate ontology to guide the default inference (allowing rapid estimation of when one is in a case where the default assumption of location-persistence will not apply).

14.3 Temporal Logic Relationships

In principle one can do temporal inference in PLN without adding any new constructs, aside from the predicates introduced above and labeled Probabilistic Event Calculus. One can simply use standard higher-order PLN links to interrelate event calculus predicates, and carry out temporal inference in this way. However, it seems this is not the most effective approach. To do temporal inference in PLN efficiently and elegantly, it's best to introduce some new relationship types: predictive implication (PredictiveImplication and EventualPredictiveImplication) and

tial and simultaneous conjunction and disjunction. Here we introduce these concepts, and then describe their usage for inferring appropriate behaviors to control an agent in a simulation world.

14.3.1 Sequential AND

Conceptually, the basic idea of sequential AND is that

```
SeqAND (A1,   ...,  An )
```

should be "just like AND but with the additional rule that the order of the items in the sequence corresponds to the temporal order in which they occur." Similarly, SimOR and SimAND ("Sim" for Simultaneous) may be used to define parallelism within an SeqAND list; e.g., the following pseudocode for holding up a convenience store:

```
SeqAND
    enter store
    SimOR
          kill clerk
          knock clerk unconscious
    steal money
    leave store
```

However, there is some subtlety lurking beneath the surface here. The simplistic interpretations of SeqAND as "AND plus sequence" is not always adequate, as the event calculus notions introduced above reveal. Attention must be paid to the initiations and terminations of events. The basic idea of sequential AND must be decomposed into multiple notions, based on disambiguating properties we call disjointness and eventuality. Furthermore, there is some additional subtlety because the same sequential logical link types need to be applied to both terms and predicates.

Applied to terms, the definition of a basic, non-disjoint, binary SeqAND is

```
SeqAND A B <s, T>

iff

AND
    AND A B
    Initiation(B) - Initiation(A) lies in interval T
```

Basically, what this says is "*B* starts *x* seconds after *A* starts, where *x* lies in t terval *T*."

Note that in the above we use <*s*, *T*> to denote a truth-value with strength *s* and time-interval parameter *T*. For instance,

```
SeqAND <.8,(0s,120s)>
    shaving_event_33
    showering_event_43
```

means

```
AND
    AND
        shaving_event_33
        showering_event_43
    SS_Initiation(showering_event_43)-
    SS_Initiation(shaving_event_33) in [0s, 120s]
```

On the other hand, the definition of a basic *disjoint* binary SeqAND between terms is

```
DisjointSeqAND A B <s, T>

iff

AND
    AND A B
    Initiation(B) - Termination(A)
                                lies in interval T
```

Basically, what this says is "*B* starts *x* seconds after *A* finishes, where *x* lies in the interval *T*" – a notion quite different from plain old SeqAND.

EventualSeqAND and DisjointEventualSeqAND are defined similarly, but without specifying any particular time interval. For example,

```
EventualSeqAND A B <s>

iff

AND
    AND A B
    Evaluation
        after
        List
            SS_Initiation(B)
```

```
SS_Initiation(A)
```

Next, there are several natural ways to define (ordinary or disjoint) SeqAND as applied to predicates. The method we have chosen makes use of a simple variant of "situation semantics" (Barwise 1983). Consider P and Q as predicates that apply to some situation; e.g.,

```
P(S) = shave(S) =
true if shaving occurs in situation S

Q(S) = shower(S) =
true if showering occurs in situation S
```

Let

```
timeof(P,S) =
the set of times at which the predicate P is true in situation S
```

Then

```
SeqAND P Q
```

is also a predicate that applies to a situation; i.e.,

```
(SeqAND P Q)(S) <s,T>
```

is defined to be true of situation S iff

```
AND <s>
    AND P(S) Q(S)
    timeof(Q,S) - timeof(P,S) intersects interval T
```

In the case of an *n*-ary sequential AND, the time interval *T* must be replaced by a series of time intervals; e.g.,

```
SeqAND <s,(T1,…,Tn)>
    A1
    ...
    An-1
    An
```

is a shorthand for

```
AND <s>
    SeqAND A1 A2 <T1>
    ...
    SeqAND An-1 An <Tn>
```

Simultaneous conjunction and disjunction are somewhat simpler to handle. We can simply say, for instance,

```
SimAND A B <s, T>

iff

AND
    HoldsThroughout(B,T)
    HoldsThroughout(A,T)
```

and make a similar definition for SimOr. Extension from terms to predicates using situation semantics works analogously for simultaneous links as for sequential ones. Related link types ExistentialSimAND and ExistentialSimOR, defined e.g. via

```
SimOR A B <s, T>

iff

OR
    HoldsSometimeIn(B,T)
    HoldsSometimeIn(A,T)
```

may also be useful.

14.3.2 Predictive Implication

Next, having introduced the temporal versions of conjunction (and disjunction), we introduce the temporal version of implication: the relationship

```
ExtensionalPredictiveImplication P Q <s,T>
```

which is defined as

```
ExtensionalImplication <s>
    P(S)
    [SeqAND( P ,  Q) <T>](S)
```

is a related notion of DisjointPredictiveImplication defined in terms of DisjointSeqAND.
PredictiveImplication may also be meaningfully defined intensionally; i.e.,

```
IntensionalPredictiveImplication P Q <s,T>
```

may be defined as

```
IntensionalImplication <s>
    P(S)
    [SeqAND( P ,  Q) <T>](S)
```

and of course there is also mixed PredictiveImplication, which is, in fact, the most commonly useful kind.

14.3.3 Eventual Predictive Implication

Predictive implication is an important concept but applying it to certain kinds of practical situations can be awkward. It turns out to be useful to also introduce a specific relationship type with the semantics "If X continues for long enough, then Y will occur." In PLN this is called EventualPredictiveImplication, so that, e.g., we may say

```
EventualPredictiveImplication starve die

EventualPredictiveImplication run sweat
```

Formally, for events X and Y

```
EventualPredictiveImplication P Q
```

may be considered as a shorthand for

```
Implication <s>
    P(S)
    [EventualSeqAND( P ,  Q) <T>](S)
```

There are also purely extensional and intensional versions, and there is a notion of DisjointEventualPredictiveImplication as well.

14.3.4 Predictive Chains

Finally, we have found use for the notion of a predictive chain, where

```
PredictiveChain <s,(T1,…,Tn)>
    A1
    ...
    An-1
    An
```

means

```
PredictiveImplication <s,Tn>
    SeqAND <(T1,…,Tn-1)>
          A1
          ...
          An-1
    An
```

For instance,

```
PredictiveChain
    Teacher is thirsty
      I go to the water machine
    I get a cup
    I fill the cup
    I bring the cup to teacher
    Teacher is happy
```

Disjoint and eventual predictive chains may be introduced in an obvious way.

14.3.5 Inference on Temporal Relationships

Inference on temporal-logical relationships must use both the traditional truth-values and the probability of temporal precedence; e.g., in figuring out whether

```
PredictiveImplication A B
PredictiveImplication B C
|-
PredictiveImplication A C
```

one must calculate the truth-value of

```
Implication A C
```

but also the odds that in fact A occurs before C. The key point here, conceptually, is that the probabilistic framework may be applied to time intervals, allowing PLN to serve as a probabilistic temporal logic not just a probabilistic static logic.

In the context of indefinite probabilities, the use of time distributions may be viewed as adding an additional level of Monte Carlo calculation. In handling each premise in an inference, one may integrate over all time-points, weighting each one by its probability according to the premise's time distribution. This means that for each collection of (premise, time point) pairs, one does a whole inference; and then one revises the results, using the weightings of the premises' time distributions.

14.4 Application to Making a Virtual Agent Learn to Play Fetch

As an example of PLN temporal inference, we describe here experiments that were carried out using temporal PLN to control the learning engine of a humanoid virtual agent carrying out simple behaviors in a 3D simulation world. Specifically, we run through in detail how PLN temporal inference was used to enable the agent to learn to play the game of fetch in the AGISim game world.

The application described here was not a pure PLN application, but PLN's temporal inference capability lay at its core. More broadly, the application involved the Novamente Cognition Engine architecture and the application of PLN, in combination with other simpler modules, within this architecture. The NCE is extremely flexible, incorporating a variety of carefully inter-coordinated learning processes, but the experiments described in this section relied primarily on the integration of PLN inference with statistical pattern mining based perception and functional program execution based agent control.

From the PLN and Novamente point of view, the experiments reported here are interesting mostly as a "smoke test" for embodied, inference-based reinforcement learning, to indicate that the basic mechanisms required for doing this sort of inferential learning are integrated adequately and working correctly.

One final preliminary note: Learning to play fetch in the manner described here requires the assumption that the system already has learned (or been provided intrinsically with the capability for) how to recognize objects: balls, teachers, and the like. Object recognition in AGISim is something the NCE is able to learn, given the relatively low-bandwidth nature of the perceptions coming into it from AGISim, but that was not the focus of the reported work. Instead, we will take this capability for granted here and focus on perception/action/cognition integration specific to the "fetch" task.

14.4.1 The Game of Fetch

"Fetch" is a game typically played between a human and a dog. The basic idea is a simple one: the human throws an object and says "fetch," the dog runs to the object, picks it up, and brings it back to the human, who then rewards the dog for correct behavior. In our learning experiments, the teacher (a humanoid agent in AGISim) plays the role of the human and the Novamente-controlled agent plays the role of the dog.

In more complex AGISim experiments, the teacher is actually controlled by a human being, who delivers rewards to Novamente by using the Reward controls on the AGISim user interface. Due to the simplicity of the current task, in the "fetch" experiments reported here the human controller was replaced by automated control code. The critical aspect of this automated teacher is partial reward. That is, you can't teach a dog (or a baby Novamente) to play fetch simply by rewarding it when it successfully retrieves the object and brings it to you, and rewarding it not at all otherwise – because the odds of it ever carrying out this "correct" behavior by random experimentation in the first place would be very low.. What is needed for instruction to be successful is for there to be a structure of partial rewards in place. We used here a modest approach with only one partial reward preceding the final reward for the target behavior. The partial reward is initially given for the agent when it manages to lift the ball from the ground, and once it has learnt to repeat this behavior we switch to the final stage where we only reward the agent when it proceeds to take the ball to the teacher and drop it there.

14.4.2 Perceptual Pattern Mining

Pattern mining, within the NCE, is a process that identifies frequent or otherwise significant patterns in a large body of data. The process is independent of whether the data is perceptual or related to actions, cognition, etc. However, it is often associated with perceptual data and abstractions from perceptual data. In principle, everything obtained via pattern mining could also be obtained via inference, but pattern mining has superior performance in many applications.

The pattern mining relevant to the fetch learning consists of simply recognizing frequent sequences of events such as

```
SequentialAND
    SimultaneousAND
       I am holding the ball
       I am near the teacher
     I get more reward.
```

in full-fledged PLN notation is

```
SequentialAND
  SimultaneousAND
    EvaluationLink
      holding
      ball
    EvaluationLink
      near
      ListLink (me, teacher)
  Reward
```

14.4.3 Particularities of PLN as Applied to Learning to Play Fetch

From a PLN perspective, learning to play fetch is a simple instance of backward chaining inference – a standard inference control mechanism whose customization to the PLN context was briefly described in the previous chapter. The goal of the NCE in this context is to maximize reward, and the goal of the PLN backward chainer is to find some way to prove that if some actionable predicates become true, then the truth-value of

```
EvaluationLink(Reward)
```

is maximized. This inference is facilitated by assuming that any action can be tried out; i.e., the trying of actions is considered to be in the axiom set of the inference. (An alternative, also workable approach is to set a PredictiveImplicationLink($1, Reward) as the target of the inference, and launch the inference to fill in the variable slot $1 with a sequence of actions.)

PredictiveImplicationLink is a Novamente Link type that combines logical (probabilistic) implication with temporal precedence. Basically, the backward chainer is being asked to construct an Atom that implies reward in the future. Each PredictiveImplicationLink contains a time-distribution indicating how long the target is supposed to occur after the source does; in this case the time-distribution must be centered on the rough length of time that a single episode of the "fetch" game occupies.

To learn how to play fetch, The NCE must repeatedly invoke PLN backward chaining on a knowledge base consisting of Atoms that are constantly being acted upon by perceptual pattern mining as discussed above. PLN learns logical knowledge about circumstances that imply reward, and then a straightforward process called predicate schematization produces NCE objects called "executable schemata," which are then executed. This causes the system to carry out actions, which in turn lead to new perceptions, which give PLN more information to guide its reasoning and lead to the construction of new procedures, etc.

In order to carry out very simple inferences about schema execution as required in the fetch example, PLN uses two primitive predicates:

- **try(X),** indicating that the schema X is executed
- **can(X),** indicating that the necessary preconditions of schema X are fulfilled, so that the successful execution of X will be possible

Furthermore, the following piece of knowledge is assumed to be known by the system, and is provided to the NCE as an axiom:

```
PredictiveImplication
    SimultaneousAnd
        Evaluation try X
        Evaluation can X
    Evaluation done X
```

This simply means that if the system can do X, and tries to do X, then at some later point in time, it has done X. Note that this implication may also be used probabilistically in cases where it is not certain whether or not the system can do X. Note that in essentially every case, the truth value of `Evaluation can X` will be uncertain (even if an action is extremely simple, there's always some possibility of an error message from the actuator), so that the output of this PredictiveImplication will essentially never be crisp.

The proper use of the "can" predicate necessitates that we mine the history of occasions in which a certain action succeeded and occasions in which it did not. This allows us to create PredictiveImplications that embody the knowledge of the preconditions for successfully carrying out the action. In the inference experiments reported here, we use a simpler approach because the basic mining problem is so easy; we just assume that "can" holds for all actions, and push the statistics of success/failure into the respective truth-values of the PredictiveImplications produced by pattern mining.

Next, we must explain a few shorthands and peculiarities that we introduced when adapting the PLN rules to carry out the temporal inference required in the fetch example. The ModusPonensRule used here is simply a probabilistic version of the standard Boolean modus ponens, as described earlier. It can also be applied to PredictiveImplications, insofar as the system keeps track of the structure of the proof tree so as to maintain the (temporally) proper order of arguments. By following the convention that the order in which the arguments to modus ponens must be applied is always the same as the related temporal order, we may extract a plan of consecutive actions, in an unambiguous order, directly from the proof tree.

Relatedly, the AndRules used are of the form

B
|-
A & B

and can be supplied with a temporal inference formula so as to make them applicable for creating SequentialANDs. We will also make use of what we call the SimpleANDRule, which embodies a simplistic independence assumption and finds the truth-value of a whole conjunction based only on the truth-values of its individual constituents, without trying to take advantage of the truth-values possibly known to hold for partial conjunctions.

We use a "macro rule" called RewritingRule, which is defined as a composition of AndRule and ModusPonensRule. It is used as a shorthand for converting atoms from one form to another when we have a Boolean true implication at our disposal.

What we call CrispUnificationRule is a "bookkeeping rule" that serves simply to produce, from a variable-laden expression (Atom defined by a ForAll, ThereExists or VariableScope relationship), a version in which one or more variables have been bound. The truth-value of the resulting atom is the same as that of the quantified expression itself.

Finally, we define the specific predicates used as primitives for this learning experiment, which enables us to abstract away from any actual motor learning:

- **Reward** – a built-in sensation corresponding to the Novamente agent getting Reward, either via the AGISim teaching interface or otherwise having its internal Reward indicator stimulated
- **goto** – a persistent event; goto(x) means the agent is going to x
- **lift** – an action; lift(x) means the agent lifts x
- **drop** – an action; drop(x) means the agent drops x if it is currently holding x (and when this happens close to an agent T, we can interpret that informally as "giving" x to T)
- **TeacherSay** – a percept; TeacherSay(x) means that the teacher utters the string x
- **holding** – a persistent event; holding(x) means the agent is holding x

14.4.4 Learning to Play Fetch Via PLN Backward Chaining

Next, we show a PLN inference trajectory that results in learning to play fetch once we have proceeded into the final reward stage. This trajectory is one of many produced by PLN in various indeterministic learning runs. When acted upon by the NCE's predicate schematization process, the conclusion of this trajectory (depicted graphically in Figure 1) produces the simple schema (executable procedure)

```
try goto Ball
try lift Ball
try goto Teacher
try drop Ball.
```

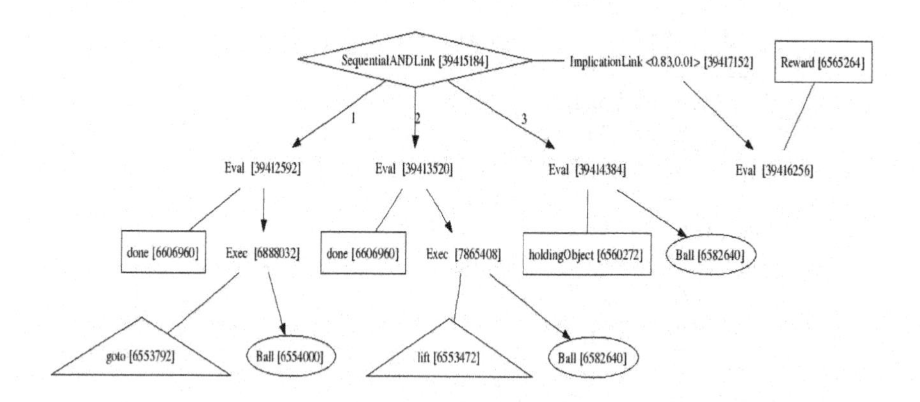

Figure 1. Graphical depiction of the final logical plan learned for carrying out the fetch task

It is quite striking to see how much work PLN and the perception system need to go through to get to this relatively simple plan, resulting in an even simpler logical procedure! Nevertheless, the computational work required to do this sort of inference is quite minimal, and the key point is that the inference is done by a general-purpose inference engine that was not at all tailored for this particular task. The inference target was

```
EvaluationLink <0.80, 0.0099>
    Reward: PredicateNode <1, 0>.
```

The final truth-value found for the EvaluationLink is of the form <strength, weight of evidence>, meaning that the inference process initially found a way to achieve the Reward with a strength of 0.80, but with a weight of evidence of only .0099 (the rather unforgiving scaling factor of which originates from the internals of the perception pattern miner). Continuing the run makes the strength and weight of evidence increase toward 1.0.

The numbers such as [9053948] following nodes and links indicate the "handle" of the entity in the NCE's knowledge store, and serve as unique identifiers. The target was produced by applying ModusPonensRule to the combination of

```
PredictiveImplicationLink <0.8,0.01> [9053948]
  SequentialAndLink <1,0> [9053937]
    EvaluationLink <1,0> [905394208]
      "holdingObject":PredicateNode <0,0> [6560272]
      ListLink [7890576]
        "Ball":ConceptNode <0,0> [6582640]
    EvaluationLink <1,0> [905389520]
      "done":PredicateNode <0,0> [6606960]
      ListLink [6873008]
        ExecutionLink [6888032]
          "goto":GroundedSchemaNode <0,0> [6553792]
          ListLink [6932096]
            "Teacher":ConceptNode <0,0> [6554000]
    EvaluationLink <1,0> [905393440]
      "try":PredicateNode <0,0> [6552272]
      ListLink [7504864]
        ExecutionLink [7505792]
          "drop":GroundedSchemaNode <0,0> [6564640]
           ListLink [7506928]
            "Ball":ConceptNode <0,0> [6559856]
  EvaluationLink <1,0> [905391056]
    "Reward":PredicateNode <1,0> [191]
```

(the plan fed to predicate schematization, and shown in Figure 1) and

```
SequentialAndLink <1,0.01> [840895904]
  EvaluationLink <1,0.01> [104300720]
    "holdingObject":PredicateNode <0,0> [6560272]
    ListLink [7890576]
      "Ball":ConceptNode <0,0> [6582640]
  EvaluationLink <1,1> [72895584]
    "done":PredicateNode <0,0> [6606960]
    ListLink [6873008]
      ExecutionLink [6888032]
        "goto":GroundedSchemaNode <0,0> [6553792]
        ListLink [6932096]
          "Teacher":ConceptNode <0,0> [6554000]
  EvaluationLink <1,1> [104537344]
    "try":PredicateNode <0,0> [6552272]
    ListLink [7504864]
      ExecutionLink [7505792]
        "drop":GroundedSchemaNode <0,0> [6564640]
        ListLink [7506928]
          "Ball":ConceptNode <0,0> [6559856]
```

Next, the SequentialANDLink [840895904] was produced by applying Si ANDRule to its three child EvaluationLinks.

The EvaluationLink [104300720] was produced by applying ModusPonensRule to

```
PredictiveImplicationLink <1,0.01> [39405248]
  SequentialANDLink <1,0> [39403472]
    EvaluationLink <1,0> [39371040]
      "done":PredicateNode <0,0> [6606960]
     ListLink [7511296]
       ExecutionLink [7490272]
         "goto":GroundedSchemaNode<0,0> [6553792]
         ListLink [7554976]
           "Ball":ConceptNode <0,0> [6558784]
   EvaluationLink <1,0> [39402640]
    "done":PredicateNode <0,0> [6606960]
    ListLink [7851408]
      ExecutionLink [7865376]
        "lift":GroundedSchemaNode <0,0> [6553472]
        ListLink [7890576]
          "Ball":ConceptNode <0,0> [6582640]
 EvaluationLink <1,0> [39404448]
   "holdingObject":PredicateNode <0,0> [6560272]
   ListLink [7890576]
     "Ball":ConceptNode <0,0> [6582640]
```

which was mined from perception data (which includes proprioceptive observation data indicating that the agent has completed an elementary action), and to

```
SequentialANDLink <1,1> [104307776]
   EvaluationLink <1,1> [72926800]
      "done":PredicateNode <0,0> [6606960]
      ListLink [7511296]
         ExecutionLink [7490272]
            "goto":GroundedSchemaNode <0,0> [6553792]
            ListLink [7554976]
               "Ball":ConceptNode <0,0> [6558784]
   EvaluationLink <1,1> [72913264]
      "done":PredicateNode <0,0> [6606960]
      ListLink [7851408]
         ExecutionLink [7865376]
            "lift":GroundedSchemaNode <0,0> [6553472]
            ListLink [7890576]
               "Ball":ConceptNode <0,0> [6582640].
```

The SequentialANDLink [104307776] was produced by applying SimpleANDRule to its two child EvaluationLinks. The EvaluationLink [72926800] was produced by applying RewritingRule to

```
EvaluationLink <1,1> [72916304]
   "try":PredicateNode <0,0> [6552272]
   ListLink [7511296]
      ExecutionLink [7490272]
         "goto":GroundedSchemaNode <0,0> [6553792]
         ListLink [7554976]
            "Ball":ConceptNode <0,0> [6558784]
```

and

```
EvaluationLink <1,1> [72923504]
   "can":PredicateNode <1,0> [6566128]
   ListLink [7511296]
      ExecutionLink [7490272]
         "goto":GroundedSchemaNode <0,0> [6553792]
         ListLink [7554976]
            "Ball":ConceptNode <0,0> [6558784].
```

The EvaluationLink [72916304], as well as all other *try* statements, were considered axiomatic and technically produced by applying the CrispUnificationRule to

```
ForallLink <1,1> [6579808]
   ListLink <1,0> [6564144]
      "$A":VariableNode <1,0> [6563968]
   EvaluationLink <1,0> [6579200]
      "try":PredicateNode <0,0> [6552272]
      ListLink <1,0> [6564144]
         "$A":VariableNode <1,0> [6563968]
```

The EvaluationLink [72923504], as well as all other *can* statements, were considered axiomatic and technically produced by applying CrispUnificationRule to:

```
ForallLink <1,1> [6559424]
   ListLink <1,0> [6564496]
      "$B":VariableNode <1,0> [6564384]
   EvaluationLink <1,0> [6550720]
      "can":PredicateNode <1,0> [6566128]
      ListLink <1,0> [6564496]
         "$B":VariableNode <1,0> [6564384]
```

The EvaluationLink [72913264] was produced by applying RewritingRule

```
EvaluationLink <1,1> [72903504]
   "try":PredicateNode <0,0> [6552272]
   ListLink [7851408]
      ExecutionLink [7865376]
         "lift":GroundedSchemaNode <0,0> [6553472]
         ListLink [7890576]
            "Ball":ConceptNode <0,0> [6582640]
```

and

```
EvaluationLink <1,1> [72909968]
   "can":PredicateNode <1,0> [6566128]
   ListLink [7851408]
      ExecutionLink [7865376]
         "lift":GroundedSchemaNode <0,0> [6553472]
         ListLink [7890576]
            "Ball":ConceptNode <0,0> [6582640]
```

And finally, returning to the first PredictiveImplicationLink's children, EvaluationLink [72895584] was produced by applying RewritingRule to the axioms

```
EvaluationLink <1,1> [72882160]
   "try":PredicateNode <0,0> [6552272]
   ListLink [6873008]
      ExecutionLink [6888032]
         "goto":GroundedSchemaNode <0,0> [6553792]
         ListLink [6932096]
            "Teacher":ConceptNode <0,0> [6554000]
```

and

```
EvaluationLink <1,1> [72888224]
   "can":PredicateNode <1,0> [6566128]
   ListLink [6873008]
      ExecutionLink [6888032]
         "goto":GroundedSchemaNode <0,0> [6553792]
         ListLink [6932096]
            "Teacher":ConceptNode <0,0> [6554000]
```

In conclusion, we have given a relatively detailed treatment of a simple learning experiment – learning to play fetch – conducted with the NCE integrative AI system in the AGISim simulation world. Our approach was to first build an integrative AI architecture we believe to be capable of highly general learning, and en apply it to the fetch test, while making minimal parameter adjustment to the specifics of the learning problem. This means that in learning to play fetch the system has to deal with perception, action, and cognition modules that are not fetch-specific, but are rather intended to be powerful enough to deal with a wide variety of learning tasks corresponding to the full range of levels of cognitive development.

Ultimately, in a problem this simple the general-intelligence infrastructure of the NCE and the broad sophistication of PLN don't add all that much. There exist much simpler systems with equal fetch-playing prowess. For instance, the PLN system's capability of powerful analogical reasoning is not being used at all here, and its use in an embodiment context is a topic for another paper. However, this sort of simple integrated learning lays the foundation for more complex embodied learning based on integrated cognition, the focus of much of our ongoing work.

14.5 Causal Inference

Temporal inference, as we have seen, is relatively conceptually simple from a probabilistic perspective. It leads to a number of new link types and a fair amount of bookkeeping complication (the node-and-link constructs shown in the context of the fetch example won't win any prizes for elegance), but is not fundamentally conceptually problematic. The tricky issues that arise, such as the frame problem, are really more basic AI issues than temporal inference issues in particular.

Next, what about causality? This turns out to be a much subtler matter. There is much evidence that human causal inference is pragmatic and heterogeneous rather than purely mathematical (see discussion and references in Goertzel 2006). One illustration of this is the huge variance in the concept of causality that exists among various humans and human groups (Smith 2003). Given this, it's not to be expected that PLN or any other logical framework could, in itself, give a thorough foundation for understanding causality. But even so, there are interesting connections to be drawn between PLN and aspects of causal inference.

Predictive implication, as discussed above, allows us to discuss temporal correlation in a pragmatic way. But this brings us to what is perhaps the single most key conceptual point regarding causation: *correlation and causation are distinct*. To take the classic example, if a rooster regularly crows before dawn, we do not want to infer that he causes the sun to rise.

In general, if X appears to cause Y, it may actually be due to Z causing both X and Y, with Y appearing later than X. We can only be sure that this is not the case if we have a way to identify alternative causes and test them in comparison to the causes we think are real. Or, as in the rooster/dawn case, we may have background knowledge that makes the "X causes Y" scenario intrinsically implausible in terms of the existence of potential causal mechanisms.

Let's consider this example in a little more detail. In the case of rooster dawn, clearly we have both implication and temporal precedence. Hence there will be a PredictiveImplication from "rooster crows" to "sun rises." But will the reasoning system conclude from this PredictiveImplication that if a rooster happens to crow at 1 AM the sun is going to rise really early that morning – say, at 2 AM? How is this elementary error avoided?

There are a couple of answers here. The first has to do with the intension/extension distinction. It says: *The strength of this particular PredictiveImplication may be set high by direct observation, but it will be drastically lowered by inference from more general background knowledge*. Specifically, much of this inference will be *intensional* in nature, as opposed to the purely extensional information (direct evidence-counting) that is used to conclude that roosters crowing imply sun rising. We thus conclude that one signifier of bogus causal relationships is when

```
ExtensionalPredictiveImplication A B
```

has a high strength but

```
IntensionalPredictiveImplication A B
```

has a low strength. In the case of

```
A = rooster crows
B = sun rises
```

the weight-of-evidence of the intensional relationship is much higher than that of the extensional relationship, so that the overall PredictiveImplication relationship comes out with a fairly low strength.

To put it more concretely, if the reasoning system had never seen roosters crow except an hour before sunrise, and had never seen the sun rise except after rooster crowing, the posited causal relation might indeed be created. What would keep it from surviving for long would be some knowledge about the mechanisms underlying sunrise. If the system knows that the sun is very large and rooster crows are physically insignificant forces, then this tells it that there are many possible contexts in which rooster crows would not precede the sun rising. Conjectural reasoning about these possible contexts leads to negative evidence in favor of the implication

```
PredictiveImplication rooster_crows sun_rises
```

which counterbalances – probably overwhelmingly – the positive evidence in favor of this relationship derived from empirical observation.

More concretely, one has the following pieces of evidence:

```
PredictiveImplication <.00, .99>
   small_physical_force
   movement_of_large_object
PredictiveImplication <.99,.99>
   rooster_crows
      small_physical_force
PredictiveImplication <.99, .99>
   sun_rises
      movement_of_large_object
PredictiveImplication <.00,.99>
   rooster_crows
   sun_rises
```

which must be merged with

```
PredictiveImplication rooster_crows sun_rises  <1,c>
```

derived from direct observation. So it all comes down to: How much more confident is the system that a small force can't move a large object, than that rooster crows always precede the sunrise? How big is the parameter we've denoted *c* compared to the confidence we've arbitrarily set at .99?

Of course, for this illustrative example we've chosen only one of many general world-facts that contradicts the hypothesis that rooster crows cause the sunrise... in reality many, many such facts combine to effect this contradiction. This simple example just illustrates the general point that reasoning can invoke background knowledge to contradict the simplistic "correlation implies causation" conclusions that sometimes arise from direct empirical observation.

14.5.1 Aspects of Causality Missed by a Purely Logical Analysis

In this section we will briefly discuss a couple of aspects of causal inference that seem to go beyond pure probabilistic logic – and yet are fairly easily integrable into a PLN-based framework. This sort of discussion highlights what we feel will ultimately be the greatest value of the PLN formalism; it formulates logical inference in a way that fits in naturally with a coherent overall picture of cognitive function. Here we will content ourselves with a very brief sketch of these ideas, as to pursue it further would lead us too far afield.

14.5.1.1 Simplicity of Causal Mechanisms

The first idea we propose has to do with the notion of causal mechanism. The basic idea is, given a potential cause-effect pair, to seek a concrete function mapping the cause to the effect, and to consider the causality as more substantial if this function is simpler. In PLN terms, this means that one is not only looking at the

IntensionalPredictiveImplication relationship underlying a posited causal rel ship, but one is weighting the count of this relationship more highly if the Predicates involved in deriving the relationship are simpler. This heuristic for count-biasing means that one is valuing simple causal mechanisms as opposed to complex ones. The subtlety lies in the definition of the "simplicity" of a predicate, which relies on pattern theory (Goertzel 2006) as introduced above in the context of intensional inference.

14.5.1.2 Distal Causes, Enabling Conditions

As another indication of the aspects of the human judgment of causality that are omitted by a purely logical analysis, consider the distinction between *local* and *distal* causes. For example, does an announcement by Greenspan cause the market to change, or is he just responding to changed economic conditions on interest rates, and they are the ultimate cause? Or, to take another example, suppose a man named Bill drops a stone, breaking a car windshield. Do we want to blame (assign causal status to) Bill for dropping the stone that broke the car windshield, or his act of releasing the stone, or perhaps the anger behind his action, or his childhood mistreatment by the owner of the car, or even the law of gravity pulling the rock down? Most commonly we would cite Bill as the cause because he was a free agent. But different causal ascriptions will be optimal in different contexts: typically, childhood mistreatment would be a mitigating factor in legal proceedings in such a case.

Related to this is the distinction between causes and so-called *enabling conditions*. Enabling conditions predictively imply their "effect," but they display no significant variation within the context considered pertinent. For, example oxygen is necessary to use a match to start a fire, but because it is normally always present we usually ignore it as a cause, and it would be called an enabling condition. If it really is always present, we can ignore it in practice; the problem occurs when it is very often present but sometimes is not, as for example when new unforeseen conditions occur.

We believe it is fairly straightforward to explain phenomena like distal causes and enabling conditions, but only at the cost of introducing some notions that exist in Novamente but not in PLN proper. In Novamente, Atoms are associated with quantitative "importance" values as well as truth-values. The importance value of an Atom has to do with how likely it is estimated to be that this Atom will be useful to the system in the future. There are short- and long-term importance values associated with different future time horizons. Importance may be assessed via PLN inference, but this is PLN inference based regarding propositions about how useful a given Atom has been over a given time interval.

It seems that the difference between a cause and an enabling condition often has to do with nonlogical factors. For instance, in Novamente PLN Atoms are associated not only with truth-values but also with other numbers called attention

, including for instance “importance” values indicating the expected utility of the system to thinking about the Atom. For instance, the relationship

```
PredictiveImplication oxygen fire
```

may have a high strength and count, but it is going to have a very low importance unless the AI system in question is dealing with some cases where there is insufficient oxygen available to light fires. A similar explanation may help with the distinction between distal and local causes. Local causes are the ones associated with more important predictive implications – where importance needs to be assigned, by a reasoning system, based on inferences regarding which relationships are more likely to be useful in future inferences.

Appendix A: Comparison of PLN Rules with NARS Rules

Rule	Strength formulas
Deduction	**NARS:** $s_{AC} = \frac{s_{AB} s_{BC}}{s_{AB} + s_{BC} - s_{AB} s_{BC}}$ **PLN (Independence-assumption-based):** $s_{AC} = s_{AB} s_{BC} + \frac{(1 - s_{AB})(s_C - s_B s_{BC})}{1 - s_B}$ **PLN (1D concept-geometry-based):** $s_{AC} = \frac{s_{AB} s_{BC}}{\min(s_{AB} + s_{BC}, 1)}$
Induction	**NARS:** $s_{AC} = s_{BA}$ **PLN:** $s_{AC} = \frac{s_{BA} s_{BC} s_B}{s_A} + \left(1 - \frac{s_{BA} s_B}{s_A}\right)\left(\frac{s_C - s_B s_{BC}}{1 - s_B}\right)$
Abduction	**NARS:** $s_{AC} = s_{CB}$ **PLN:** $s_{AC} = \frac{s_{AB} s_{CB} s_C}{s_B} + \frac{s_C (1 - s_{AB})(1 - s_{CB})}{1 - s_B}$

Revision	**NARS:** $s = \dfrac{\dfrac{n_1 s_1}{1-n_1} + \dfrac{n_2 s_2}{1-n_2}}{\dfrac{n_1}{1-n_1} + \dfrac{n_2}{1-n_2}}$ **PLN:** $s = \dfrac{n_1}{n_1+n_2} s_1 + \dfrac{n_2}{n_1+n_2} s_2$

A.1 Graphical Comparisons of Deduction Rules

We graphically illustrate here how the three deduction rules compare. The most similar sets of rules are those of NARS and 1-dimensional (1D) concept-geometry (Recall that in 1D concept-geometry, we consider model sets as 1-spheres on the surface of a 2-sphere.) We first depict a 3D graph of the NARS rule, for sAB and sBC, followed by the corresponding graph for 1D concept-geometry. We generated all graphs in this appendix with Mathematica.

Plot3D[sAB*sBC/(sAB + sBC - sAB*sBC), {sAB, 0, 1}, {sBC, 0, 1}]

NARS Deduction Rule

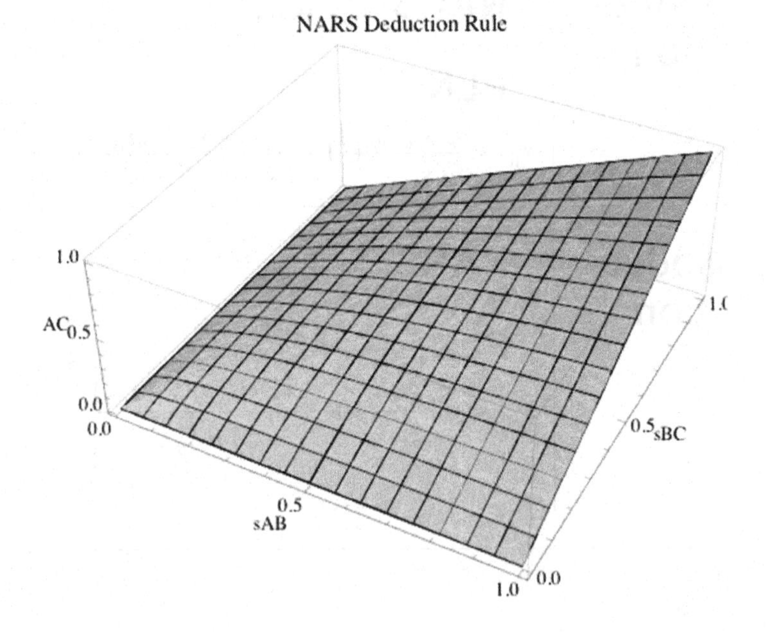

Plot3D[sAB*sBC/(Min[sAB + sBC, 1]), {sAB, 0, 1}, {sBC, 0, 1}]

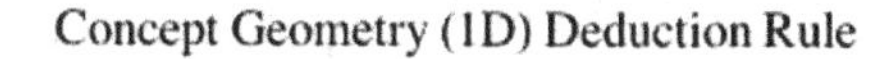

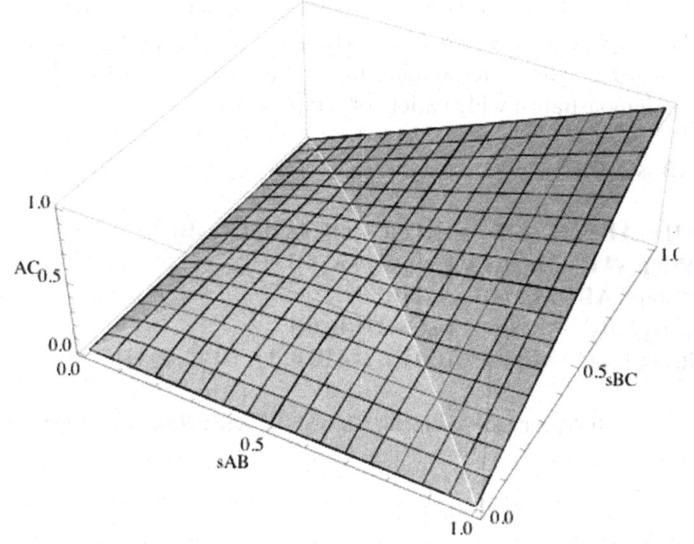

To better see the differences in these two formulas, we now show a graph of the difference between the two.

Plot3D[sAB*sBC/(sAB + sBC - sAB*sBC) - sAB*sBC/(Min[sAB + sBC, 1]), {sAB, 0, 1}, {sBC, 0, 1}]

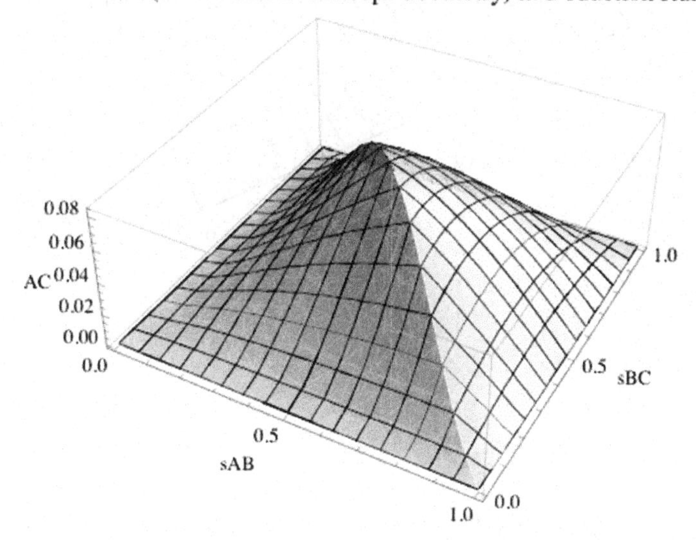

NARS and 1D concept geometry provide very similar deduction rules, with the greatest difference occurring when both sAB and sBC are close to 0.5.

We next provide graphs of the PLN independence-assumption-based deduction rule. Since this rule requires five independent variable (sA, sB, sC, sAB, sBC) rather than only the two required by NARS and 1D concept-geometry, we provide a series of 3D graphs corresponding to various input values for sA, sB, and sC, chosen to demonstrate a wide variety of graph shapes.

sA=sB=sC=0.1

```
Plot3D[sAB*sBC + (1 - sAB) (sC - sB*sAB)/(1 - sB)*
(UnitStep[sAB - Max[(sA + sB - 1)/sA, 0]] -
 UnitStep[sAB - Min[sB/sA, 1]])*
(UnitStep[sBC - Max[(sB + sC - 1)/sB, 0]] -
 UnitStep[sBC - Min[sC/sB, 1]]), {sAB, 0, 1}, {sBC, 0, 1}]
```

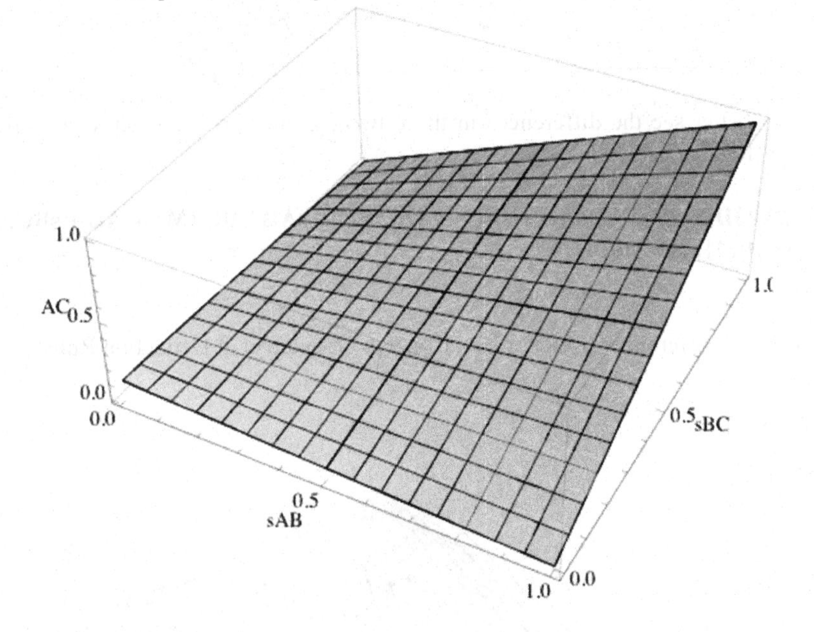

sA=sB=0.1 and sC=0.9

```
Plot3D[sAB*sBC + (1 - sAB) (sC - sB*sAB)/(1 - sB)*(UnitStep[sAB - Max[(sA + sB - 1)/sA, 0]] - UnitStep[sAB - Min[sB/sA, 1]])*(UnitStep[sBC - Max[(sB + sC - 1)/sB, 0]] - UnitStep[sBC - Min[sC/sB, 1]]), {sAB, 0, 1}, {sBC, 0, 1}]
```

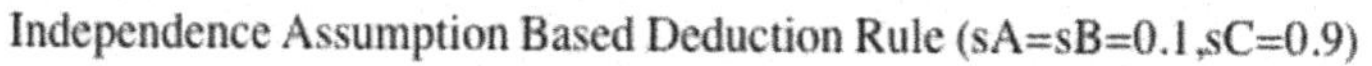

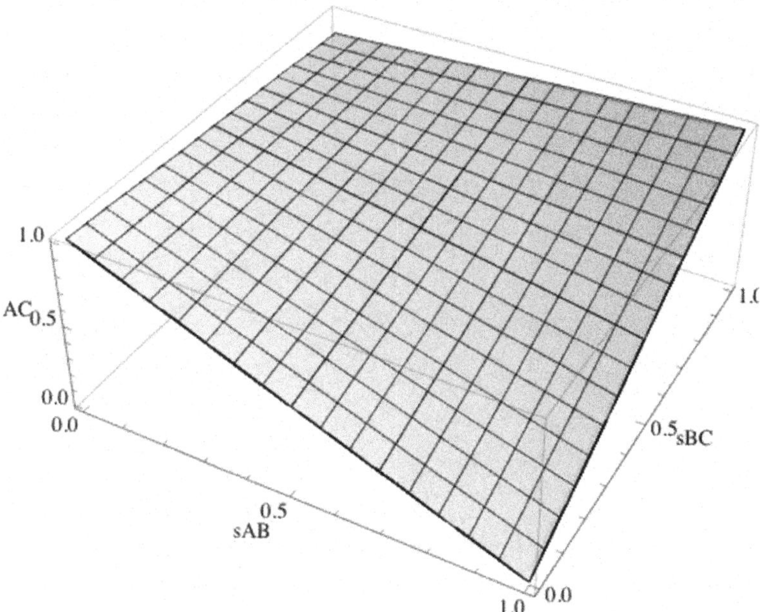

sA=sB=sC=0.9

Plot3D[sAB*sBC + (1 - sAB) (sC - sB*sAB)/(1 - sB)*(UnitStep[sAB - Max[(sA + sB - 1)/sA, 0]] - UnitStep[sAB - Min[sB/sA, 1]])*(UnitStep[sBC - Max[(sB + sC - 1)/sB, 0]] - UnitStep[sBC - Min[sC/sB, 1]]), {sAB, 0, 1}, {sBC, 0, 1}]

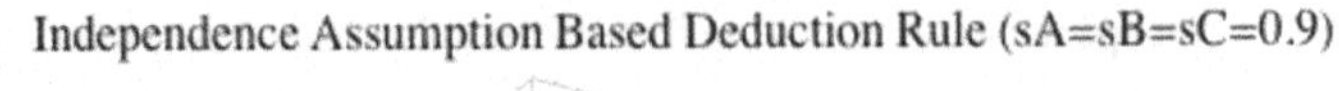

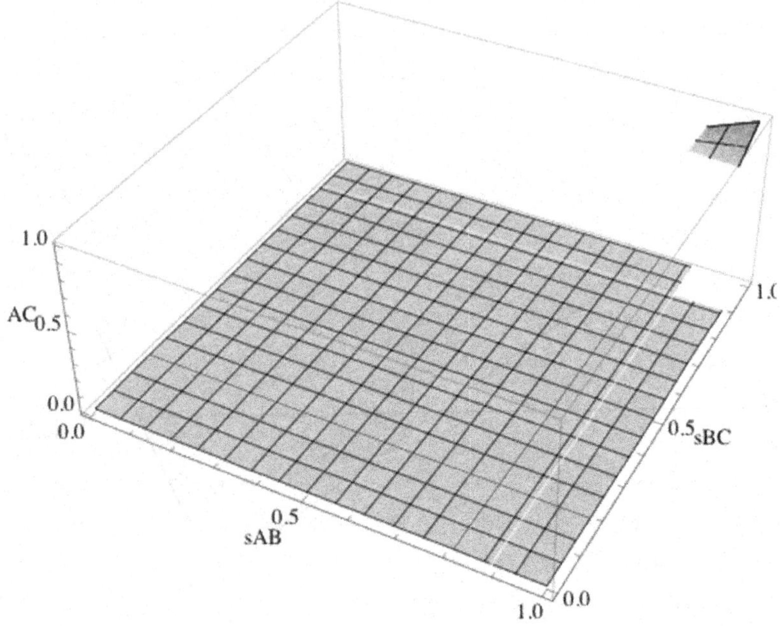

Since NARS and 1D concept-geometry deduction rules provide similar values, we provide graphs only of the differences between the NARS deduction rule and the corresponding PLN independence based rule. Once again, we provide graphs for several sets of inputs for sA, sB, and sC.

sA=sB=sC=0.1

Plot3D[sAB*sBC/(sAB + sBC - sAB*sBC) - (sAB*sBC + (1 - sAB) (sC - sB*sBC)/(1 - sB))*(UnitStep[sAB - Max[(sA + sB - 1)/sA, 0]] - UnitStep[sAB - Min[sB/sA, 1]])*(UnitStep[sBC - Max[(sB + sC - 1)/sB, 0]] - UnitStep[sBC - Min[sC/sB, 1]]), {sAB, 0, 1}, {sBC, 0, 1}]

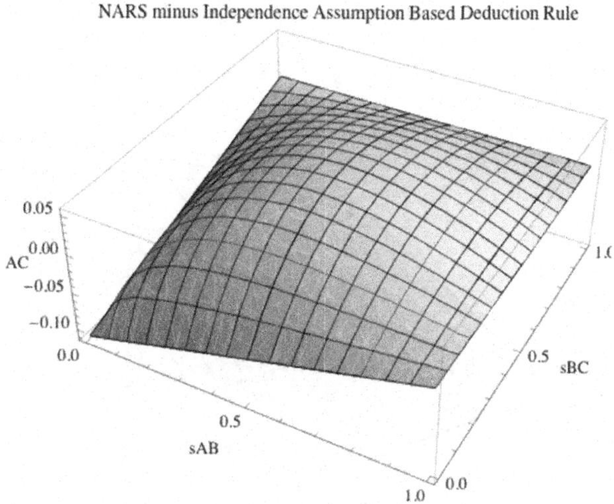

sA=sB=0.1 and sC=0.9

Plot3D[sAB*sBC/(sAB + sBC - sAB*sBC) - (sAB*sBC + (1 - sAB) (sC - sB*sBC)/(1 - sB))*(UnitStep[sAB - Max[(sA + sB - 1)/sA, 0]] - UnitStep[sAB - Min[sB/sA, 1]])*(UnitStep[sBC - Max[(sB + sC - 1)/sB, 0]] - UnitStep[sBC - Min[sC/sB, 1]]), {sAB, 0, 1}, {sBC, 0, 1}]

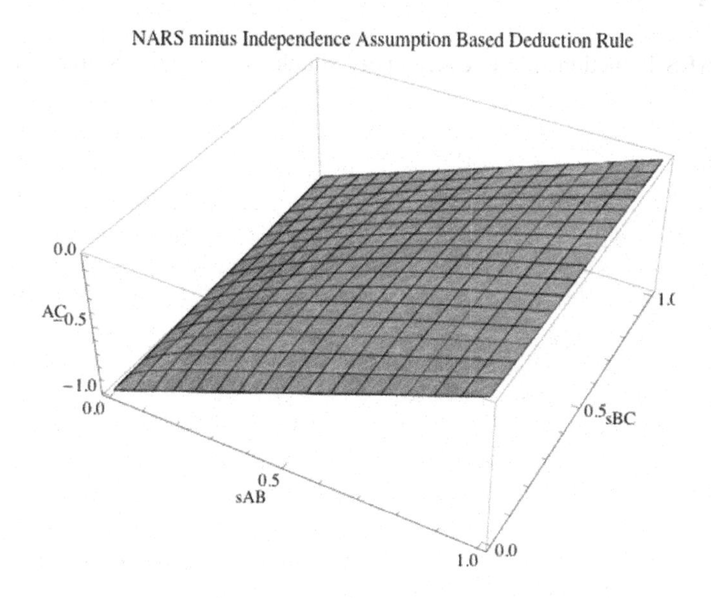

sA=sB=sC=0.9

Plot3D[sAB*sBC/(sAB + sBC - sAB*sBC) - (sAB*sBC + (1 - sAB) (sC - sB*sBC)/(1 - sB))*(UnitStep[sAB - Max[(sA + sB - 1)/sA, 0]] - UnitStep[sAB - Min[sB/sA, 1]])*(UnitStep[sBC - Max[(sB + sC - 1)/sB, 0]] - UnitStep[sBC - Min[sC/sB, 1]]), {sAB, 0, 1}, {sBC, 0, 1}]

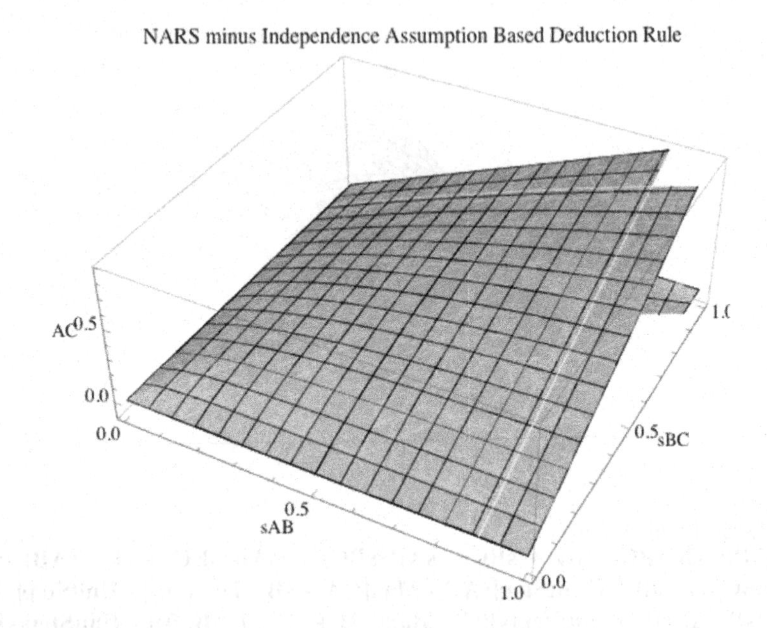

A.2 Graphical Comparisons of Induction Rules

Since the NARS Induction rule is exceedingly simple, sAC=sBA, its graph is not surprising:

Plot3D[sBA, {sBA, 0, 1}, {sBC, 0, 1}]

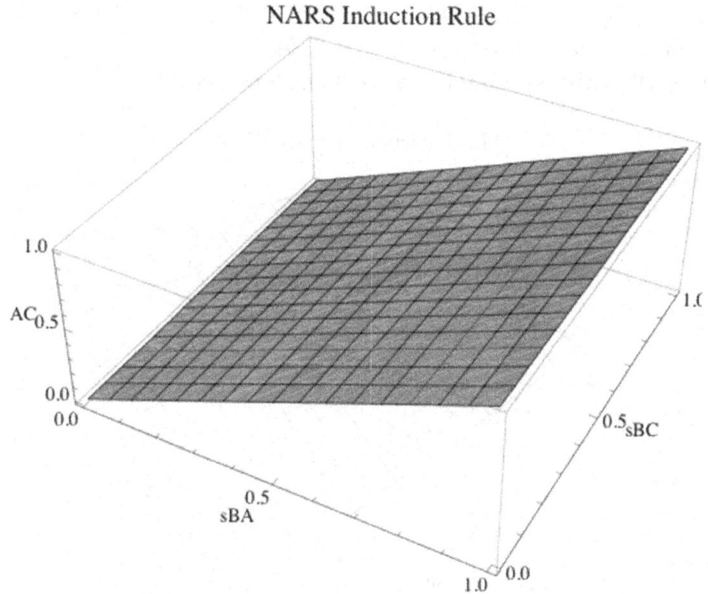

Since the PLN induction rule again has additional inputs, we provide several sets of graphs for comparison.

sA=sB=sC=0.1

Plot3D[(sBA*sBC*sB/sA + (1 - sBA*sB/sA)*(sC - sB*sBC)/(1 - sB))*(UnitStep[sBA - Max[(sB + sA - 1)/sB, 0]] - UnitStep[sBA - Min[sA/sB, 1]])*(UnitStep[sBC - Max[(sB + sC - 1)/sB, 0]] - UnitStep[sBC - Min[sC/sB, 1]]), {sBA, 0, 1}, {sBC, 0, 1}]

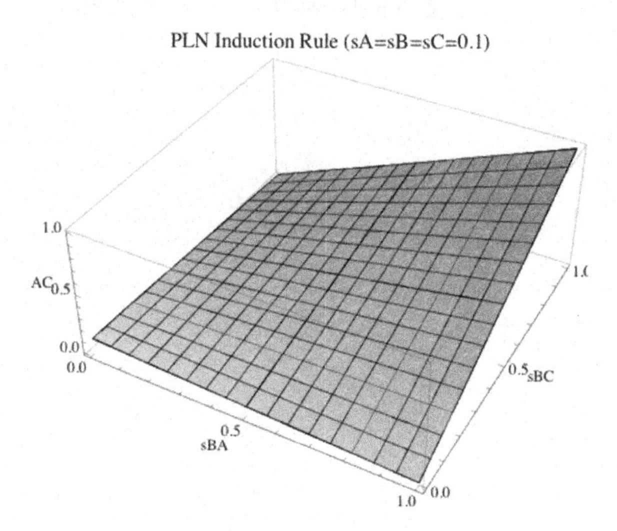

sA=sB=0.1 and sC=0.9

Plot3D[(sBA*sBC*sB/sA + (1 - sBA*sB/sA)*(sC - sB*sBC)/(1 - sB))*(UnitStep[sBA - Max[(sB + sA - 1)/sB, 0]] - UnitStep[sBA - Min[sA/sB, 1]])*(UnitStep[sBC - Max[(sB + sC - 1)/sB, 0]] - UnitStep[sBC - Min[sC/sB, 1]]), {sBA, 0, 1}, {sBC, 0, 1}]

PLN Induction Rule (sA=sB=0.1, sC=0.9)

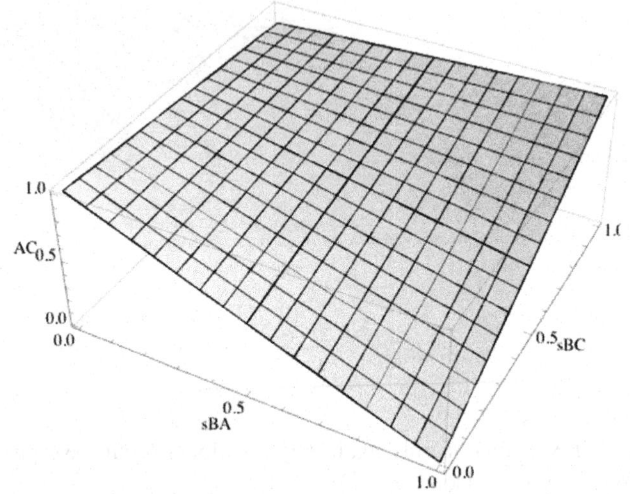

sA=sB=sC=0.9

Plot3D[(sBA*sBC*sB/sA + (1 - sBA*sB/sA)*(sC - sB*sBC)/(1 - sB))*(UnitStep[sBA - Max[(sB + sA - 1)/sB, 0]] - UnitStep[sBA - Min[sA/sB, 1]])*(UnitStep[sBC - Max[(sB + sC - 1)/sB, 0]] - UnitStep[sBC - Min[sC/sB, 1]]), {sBA, 0, 1}, {sBC, 0, 1}]

PLN Induction Rule (sA=sB=sC=0.9)

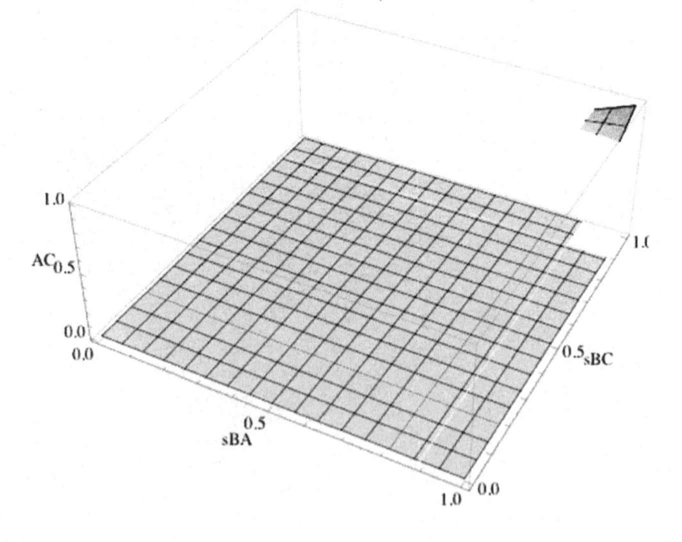

A.3Graphical Comparisons of Abduction Rules

The final rule for which we provide graphical comparisons between NARS and PLN is the abduction rule. Again, the NARS rule, sAC=sCB is very simple:

Plot3D[sCB, {sAB, 0, 1}, {sCB, 0, 1}]

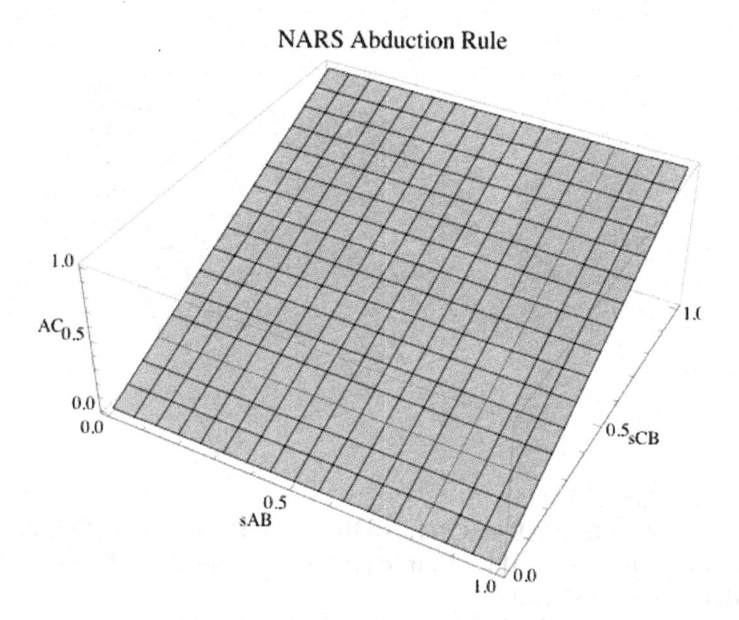

As with the other PLN rules, the PLN abduction rule requires five inputs along with consistency conditions. To compare the behavior of the PLN abduction rule with that of NARS, we again depict a series of graphs corresponding to various values for sA, sB, and sC.

sA=sB=sC=0.1

Plot3D[(sAB*sCB*sC/sB + sC*(1 - sAB)*(1 - sCB)/(1 - sB))* (UnitStep[sAB - Max[(sA + sB - 1)/sA, 0]] - UnitStep[sAB Min[sB/sA, 1]])*(UnitStep[sCB - Max[(sC + sB - 1)/sC, 0]] - UnitStep[sCB - Min[sB/sC, 1]]), {sAB, 0, 1}, {sCB, 0, 1}]

PLN Abduction Rule (sA=sB=sC=0.1)

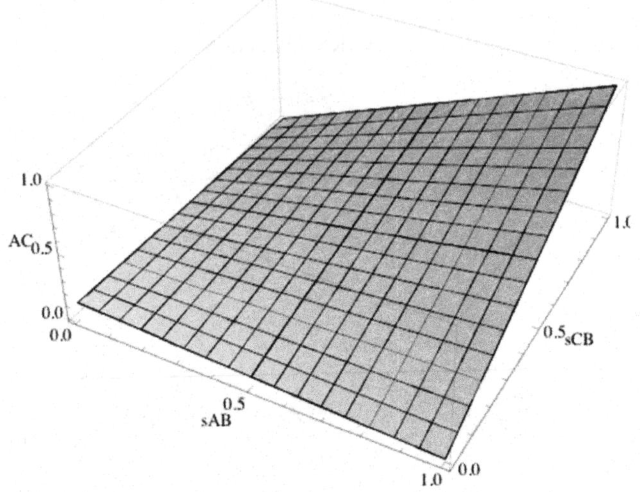

sA=sB=0.1 and sC=0.9

Plot3D[(sAB*sCB*sC/sB + sC*(1 - sAB)*(1 - sCB)/(1 - sB))* (UnitStep[sAB - Max[(sA + sB - 1)/sA, 0]] - UnitStep[sAB Min[sB/sA, 1]])*(UnitStep[sCB - Max[(sC + sB - 1)/sC, 0]] - UnitStep[sCB - Min[sB/sC, 1]]), {sAB, 0, 1}, {sCB, 0, 1}]

PLN Abduction Rule (sA=sB=0.1, sC=0.9)

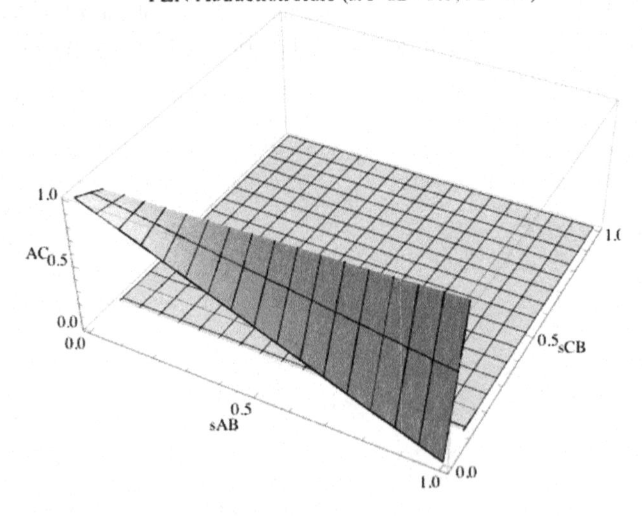

sA=sB=sC=0.9

Plot3D[(sAB*sCB*sC/sB + sC*(1 - sAB)*(1 - sCB)/(1 - sB))* (UnitStep[sAB - Max[(sA + sB - 1)/sA, 0]] - UnitStep[sAB Min[sB/sA, 1]])*(UnitStep[sCB - Max[(sC + sB - 1)/sC, 0]] - UnitStep[sCB - Min[sB/sC, 1]]), {sAB, 0, 1}, {sCB, 0, 1}]

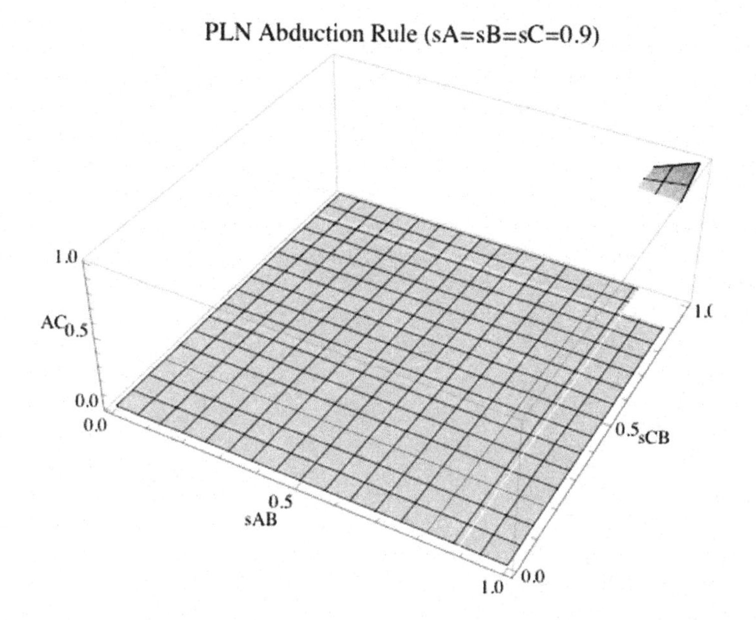

A.4A Brief Discussion of the Comparisons

For all three of deduction, induction, and abduction, it is interesting to note that the NARS and PLN appear to provide extremely similar values for the cases in which sA, sB, and sC are all small. As these values increase, PLN places additional restrictions, through the consistency conditions, upon the other possible input values. As a result, NARS and PLN values diverge as these values increase.

REFERENCES

Chapter 1 References

Bollobas B (1998) Modern graph theory. Springer-Verlag, New York

Chaitin G J (1987) Algorithmic information theory. Cambridge University Press

Cox R T (1961) Algebra of probable inference. Johns Hopkins University Press, Baltimore, MD

Curry H B, Feys R (1958). Combinatory Logic Vol. I 1. North Holland, Amsterdam

de Finetti B (1937) La prévision: ses lois logiques, ses sources subjectives, Ann. Inst. Henri Poincaré 7, 1–68. Translation reprinted in Kyburg H E, Smokler H E (eds.) (1980), Studies in subjective probability, 2nd edn (pp. 53–118). Robert Krieger, New York

Dempster, A P (1968) A generalization of Bayesian inference. Journal of the Royal Statistical Society, Series B 30:205-247

De Raedt L, Kersting K (2004) Probabilistic inductive logic programming. In: Ben-David S, Case J, Maruoka A (eds) Proceedings of the 15th International Conference on Algorithmic Learning Theory (ALT-2004), Lecture Notes in Computer Science 3244:19-36. Springer-Verlag

Goertzel B (2006) Patterns, hypergraphs and general intelligence. Proceedings of IJCNN- 2006, Workshop on human-level intelligence

Goertzel B, Pennachin C (2007) The Novamente artificial intelligence engine. In: Goertzel B, Pennachin C (eds) Artificial General Intelligence. Springer, Berlin, Heidelberg

Goertzel B, Wang P (eds) (2007) Advances in Artificial Intelligence: Concepts, architectures, and algorithms. Proceedings of the AGI Workshop 2006. IOS Press, Washington DC

Hailperin T (1996) Sentential probability logic. Origins, development, current status, and technical applications. Lehigh University Press, Bethlehem

Halpern J (2003) Reasoning about uncertainty. MIT Press, Cambridge

Hardy M, (2002) Scaled Boolean algebras. Advances in Applied Mathematics 29:243-292

Hart D, Goertzel B (2008) OpenCog: A software framework for integrative Artificial General Intelligence In: Wang P, Goertzel B, Franklin S (eds) Artificial General Intelligence 2008, Proceedings of the first AGI conference. IOS Press, Washington DC

International Symposium on Imprecise Probabilities and Their Applications (ISIPTA 2001) http://www.sipta.org/isipta01/proceedings/

International Symposium on Imprecise Probabilities and Their Applications (ISIPTA 2003) http://www.sipta.org/isipta03/proceedings/

International Symposium on Imprecise Probabilities and Their Applications (ISIPTA 2005) http://www.sipta.org/isipta05/proceedings/

International Symposium on Imprecise Probabilities and Their Applications (ISIPTA 2007) http://www.sipta.org/isipta07/proceedings/

Jaynes J (2003) Probability: The logic of science. Cambridge University Press, New York

Keynes J M (2004) A Treatise on probability. Dover Books, Mineola, NY

Murphy K P, Weiss Y, Jordan M I (1999) Loopy belief propagation for approximate inference: An empirical study. In: Proceedings of the Fifteenth Conference on Uncertainty in Artificial Intelligence

Pearl J (1988) Probabilistic reasoning in intelligent systems. Morgan-Kaufmann, San Francisco

Pearl J (2000) Causality: Models, reasoning, and inference. Cambridge University Press, New York

Peirce C S (1931-1958) Collected Papers, vols. 1-6, Hartshorne C, Weiss P. (eds); vols. 7-8, Burks A W (ed) Harvard University Press, Cambridge, MA

Shafer G (1976) A mathematical theory of evidence. Princeton University Press, Princeton, NJ

Sommers F, Englebretsen G (2000) An invitation to formal reasoning. Ashgate Pub Ltd, Aldershot, Hampshire, UK

Walley P (1991) Statistical reasoning with imprecise probabilities. Chapman and Hall, London, New York

Wang P (1996) Non-axiomatic reasoning system: Exploring the essence of intelligence. Indiana University PhD Dissertation, Indianapolis

Wang P (2001) Confidence as higher-order uncertainty. In: Proceedings of the Second International Symposium on Imprecise Probabilities and Their Applications. Ithaca, New York, 352–361

Wang P (2006) Rigid flexibility: The logic of intelligence. Springer, Dordrecht

Wang P, Goertzel B, Franklin S (eds) (2008) Artificial General Intelligence 2008, Proceedings of the first AGI conference. IOS Press, Washington, DC

Weichselberger K, Augustin T (2003) On the symbiosis of two concepts of conditional interval probability. ISIPTA 2003: 606-

Youssef S (1994) Quantum mechanics as complex probability heory. Mod.Phys.Lett A 9:2571

Zadeh L (1965) Fuzzy sets. Information Control 8:338-353

Zadeh L (1978) Fuzzy sets as a basis for a theory of possibility. Fuzzy Sets and Systems 1:3-28

Zadeh L (1989) Knowledge representation in fuzzy logic. IEEE Transactions on Knowledge and Data Engineering 1:89-100

Chapter 2 References

Goertzel B, (2006) The Hidden Pattern: A patternist philosophy of mind. Brown Walker Press

Peirce C S (1931-1958). Collected Papers, vols. 1-6. Hartshorne, C. & Weiss, P. (eds), vols. 7-8, Burks, A W (ed) Harvard University Press, Cambridge, MA

Wang P (1993) Belief revision in probability theory. The Ninth Conference of Uncertainty in Artificial Intelligence, 519-526, Washington DC, July 1993

Wang P (2004) The limitation of Bayesianism. Artificial Intelligence 158(1):97-106

Zadeh L (1989) Knowledge representation in fuzzy logic. IEEE Transactions on Knowledge and Data Engineering 1:89-100

Chapter 4 References

Cox R T (1946) Probability, frequency, and reasonable expectation. Am. Jour. Phys. 14:1-13

De Finetti B (1974-5) Theory of probability (translation by AFM Smith of 1970 book) 2 volumes. Wiley, New York

Hardy M (2002) Scaled Boolean algebras. Advances in Applied Mathematics 243-292

Keynes J M (1921, 2004) A treatise on probability. Dover Books, Mineola, NY

Ramsey F (1931) Foundations of mathematics. Londres, Routledge

Stanford Encyclopedia of Philosophy (2003) Interpretations of Probability. http://plato.stanford.edu/entries/probability-interpret/ Accessed 12 January 2008

Walley P (1991) Statistical reasoning with imprecise probabilities. Chapman and Hall, London , New York

Walley P (1996) Inferences from multinomial aata: Learning about a bag of marbles. Journal of the Royal Statistical Society. Series B (Methodological) 58:3-57

Wang P (2004) The limitation of Bayesianism. Artificial Intelligence 158(1):97-106

Weichselberger K, Augustin T (2003) On the symbiosis of two concepts of conditional interval probability. ISIPTA 2003: 606-

Chapter 5 References

Poole D (2006) Linear Algebra: A Modern Introduction, 2nd edition. Thomson Brooks/Cole, Belmont, CA

Wang P, (1993) Belief revision in probability theory. In: Proceedings of the Ninth Conference on Uncertainty in Artificial
Intelligence 519-526, Morgan Kaufmann, San Francisco

Chapter 6 References

Goertzel B, Iklé M (2008) Revision of Indefinite Probabilities via Entropy Minimization. In preparation

Kern-Isberner G, Rödder W (2004) Belief revision and information fusion on optimum entropy. International Journal of Intelligent Systems. 19:837 – 857

Chapter 7 References

Weisstein E W (2008) Pseudoinverse. MathWorld--A Wolfram Web Resource http://mathworld.wolfram.com/Pseudoinverse.html. Accessed 15 February 2008

Weisstein E W (2008) Moore-Penrose Matrix Inverse. MathWorld--A Wolfram Web Resource http://mathworld.wolfram.com/Moore-PenroseMatrixInverse.html. Accessed 15 February 2008

Chapter 8 References

Devaney R L (1989) An Introduction to Chaotic Dynamical Systems, 2nd edition. Addison-Wesley, Upper Saddle River, NJ

Langton C G (1991) Computation at the edge of chaos: Phase transitions and emergent computation. PhD thesis, The University of Michigan, Ann Arbor, MI, 1991.

Packard N H (1988) Adaptation toward the edge of chaos. In: Kelso J A S, Mandell A J, and Shlesinger M F (eds) Dynamic patterns in complex systems 293–301, World Scientific, Singapore

Chapter 9 References

Goertzel B, Pennachin C (2008) An Inferential dynamics approach to personality and emotion driven behavior determination for virtual animals. In:

Goertzel B, Pennachin C, Geissweiller N et al (2008) An integrative methodology for teaching embodied non-linguistic Agents, applied to virtual animals in Second Life. In: Wang P, Goertzel B, Franklin S (eds) Artficial General Intelligence 2008, Proceeding of the first conference on Artificial General Intelligence (AGI-08), Memphis, TN

Heckerman D (1996) A tutorial on learning with Bayesian networks. Microsoft Research tech. report MSR-TR-95-06

Holman C (1990) Elements of an expert system for determining the satisfiability of general Boolean expressions. PhD thesis Northwestern University

Looks M (2006) Competent program evolution. PhD thesis Washington University in St. Louis

Chapter 10 References

Deransart P, Ed-Dbali A, Cervoni L (1996) Prolog: The standard reference manual. Springer-Verlag, Berlin, New York

Goertzel B, Iklé M (2007) Assessing the weight of evidence implicit in an Indefinite Probability". In: 2nd International Symposium on Intelligence Computation and Applications, Wuhan, China. China University of Geosciences, Wuhan

Goertzel B, Pinto H, Heljakka A et al (2006) Using dependency parsing and probabilistic inference to extract gene/protein interactions implicit in the combination of multiple biomedical research abstracts. In: Proceedings of BioNLP-2006 Workshop at ACL-2006, New York 104-111

Iklé M, Goertzel B, Goertzel I (2007) Indefinite probabilities for general intelligence. In: Goertzel B, Wang P (eds) Advances in Artificial Intelligence: Concepts, architectures, and algorithms. Proceedings of the AGI Workshop 2006. IOS Press, Washington DC

Roark B, Sproat R (2007) Computational approaches to morphology and syntax. Oxford University Press

Spivey M (1996) An introduction to logic programming through Prolog. Prentice Hall Europe

Chapter 11 References

Iklé M, Goertzel B (2008) Probabilistic Quantifier Logic for General Intelligence: An Indefinite Probabilities Approach, In: Wang P, Goertzel B, Franklin S (eds) Artficial General Intelligence 2008, Proceeding of the first conference on Artificial General Intelligence (AGI-08), Memphis, TN

Iklé M, Goertzel B, Goertzel I (2007) Indefinite probabilities for general intelligence. In: Goertzel B, Wang P (eds) Advances in Artificial Intelligence: Concepts, architectures, and algorithms. Proceedings of the AGI Workshop 2006. IOS Press, Washington DC

Chapter 12 References

Goertzel B (1993) The structure of intelligence. Springer-Verlag, Berlin, New York

Goertzel B (1993a) The evolving mind. Gordon and Breach

Goertzel B (1997) From complexity to creativity. Plenum Press, New York

Goertzel B (2006) Patterns, hypergraphs and general intelligence. In: Proceedings of International Joint Conference on Neural Networks, IJCNN 2006 Vancouver, CA. IEEE Press

Chapter 13 References

Delgrande J, Schaub T (2003) On the relation between Reiter's default logic and its (major) variants. In: Seventh European Conference on Symbolic and Quantitative Approaches to Reasoning with Uncertainty (ECSQARU 2003) 452-463

Holman C (1990) Elements of an expert system for determining the satisfiability of general Boolean expressions. PhD thesis Northwestern University

Hudson R A (1984) Word grammar. Blackwell, Oxford

Looks M (2006) Competent program evolution. PhD thesis Washington University in St. Louis

Reiter R (1980) A logic for default reasoning. Artificial Intelligence 13(1980) 81-132

Robbins, H (1998) Some Aspects of the Sequential Design of Experiments. In: Bulletin of the American Mathematical Society, 55 527–535, 1952.

Sutton R S, Barto A G (1998) Reinforcement Learning: An Introduction (Adaptive Computation and Machine Learning
The MIT Press, Cambridge

Chapter 14 References

Barwise J, Perry J (1983) Situations and attitudes. Cambridge: MIT-Bradford

Goertzel B (2006) The hidden pattern: A patternist philosophy of mind. Brown Walker Press

Goertzel B (2006a) Patterns, hypergraphs and general intelligence. Proceedings of IJCNN- 2006, Workshop on human-level intelligence

Iklé M, Goertzel B, Goertzel I (2007) Indefinite probabilities for general intelligence. In: Goertzel B, Wang P (eds) Advances in Artificial Intelligence: Concepts, architectures, and algorithms. Proceedings of the AGI Workshop 2006. IOS Press, Washington DC

Kowalski R, Sergot M (1986) A logic-based calculus of events. New Generation Computing 4:67-95

McCarthy J (1986) Applications of circumscription to formalizing common-sense knowledge. Artificial Intelligence 28:89-116

McCarthy J, Hayes P J (1969) Some philosophical problems from the standpoint of artificial intelligence. Machine Intelligence, 4:463-502

Miller R, Shanahan M (1999) The event-calculus in classical logic – alternative axiomatizations. Electronic Transactions on Artificial Intelligence, 3(1): 77-105

Mueller E T (2006) Commonsense reasoning. Morgan Kaufmann, San Francisco

Reiter R (1991) The frame problem in the situation calculus: a simple solution (sometimes) and a completeness result for goal regression. In: Vladimir Lifschitz (ed) Artificial Intelligence and Mathematical Theory of Computation: Papers in Honor of John McCarthy: 359-380. Academic Press, New York.

Sandewall E (1998) Cognitive robotics logic and its metatheory: Features and fluents revisited. Electronic Transactions on Artificial Intelligence 2(3-4): 307-329

Smith H L (2003) Some thoughts on causation as it relates to demography and population studies. Population and Development Review 29 (3): 459–469

INDEX

2

A

B

C

D

E

F

G

H

I

J

K

L

M

N

O

P

Q

R

S

T

U

V

W

Z